COLLINS GEM
BASIC FACTS

SCIENCE

Derek McMonagle BSc Phd

Adviser
David R. Lewis BSc Phd

COLLINS
London and Glasgow

First published 1989
Reprint 10 9 8 7 6 5 4 3 2 1 0

© Wm. Collins Sons & Co. Ltd. 1989

ISBN 0 00 459108 9

Printed in Great Britain

Introduction

Basic Facts are illustrated GEM dictionaries in important school subjects, regularly updated to take account of changes in syllabus. They cover all the important ideas and topics in these subjects up to the level of first examinations.

Bold words in an entry means a word or idea developed further in a separate entry: *italic* words are highlighted for importance.

Appendixes of important physical facts, symbols, and nomenclature, together with the major classifications of the animal and plant kingdoms are also included.

Absorption When radiation travelling in one material meets the surface of a second, three things may happen to it.

(a) It may bounce back into the first material: this is called **reflection**.

(b) It may travel on in a new direction through the second material: this is called **refraction**.

(c) It may disappear into the second material: this is called absorption.

When a surface absorbs radiation the **energy** must appear in a new form. In most cases this is **heat** energy: the **temperature** increases.

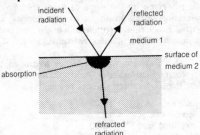

Absorption (of food) The process by which digested food particles pass from the **alimentary canal** into the bloodstream. In mammals absorption takes place in the ileum, which is part of the small intestine.

Acceleration (a) The change of **velocity**, v, of an object in unit time. The **SI unit** of acceleration is the metre per second per second, m/s^2. Acceleration can be calculated using the following equation:

$$a = \frac{v - u}{t}$$

where u is the initial velocity and v is the final velocity. Acceleration is a vector quantity. An object accelerates if its **speed** and/or its direction of motion change.

When a net outside force, F, acts on an object of mass, m, the resulting acceleration, a, is given by:

$$a = F/m$$

At any moment on a velocity/time graph the acceleration is given by the slope of the graph. Positive acceleration results in an increase in velocity while negative acceleration (often called *deceleration* or *retardation*) results in a decrease in velocity.

Acid A substance which releases **hydrogen ions** (H^+) when added to **water**. Acid solutions have a **pH** of less than 7. Common laboratory acids are:

nitric acid	HNO_3	
hydrochloric acid	HCl	strong acids
sulphuric acid	H_2SO_4	
ethanoic acid	CH_3COOH	weak acids
citric acid	$C_6H_8O_7$	

Strong acids ionize completely in water, weak acids only partially. A few acids are corrosive liquids and must be handled with care.

Acids: — turn blue **litmus** red.

— give *carbon dioxide* when added to *carbonates*.

- give **hydrogen** when added to certain **metals**.
- neutralize **alkalis**.

Acid rain Coal often contains substances which produce the acidic gases sulphur dioxide and nitrogen oxide when it is burnt. If these acidic gases are allowed to escape into the atmosphere they dissolve in any moisture present to produce acids. When the moisture falls to the earth as rain it is, in fact, acid(ic) rain.

Acid rain is responsible for considerable damage to the environment. Once soils become highly acidic, plants and animals are adversely affected and potentially harmful ions such as aluminium are able to dissolve in the acidic solution. The acidic solutions may accumulate in lakes where their effects on wild life are devastating. Recently steps have been taken to reduce the amount of acidic gases produced by British coal-fired power stations and car exhausts. In future, coal-fired power stations will have flue desulphurization units to remove acidic gases before flue gases are released into the atmosphere.

Air The **mixture** of **gases** which make up the **atmosphere** surrounding the earth. The exact composition of the air varies slightly from place to place and with height above the ground. The following diagram shows the average composition of pure air.

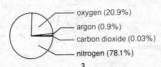

oxygen (20.9%)
argon (0.9%)
carbon dioxide (0.03%)
nitrogen (78.1%)

3

Air, particularly in industrial areas, often contains pollutants. Some common examples are shown below.

Pollutant	Sources
sulphur dioxide	burning coal and oil
carbon monoxide	engines, cigarettes
oxides of nitrogen	burning coal, car engines
soot	engines, fires
pollen	trees, plants
chlorofluorocarbons (CFCs)	aerosols

Air is vital for life. The **oxygen** it contains is necessary for **respiration**, and the **carbon dioxide** for **photosynthesis**. See **Ozone layer**.

Alimentary canal The digestive canal of an animal. In humans this canal is a tube about nine metres long running from the mouth to the anus. See **Digestion**.

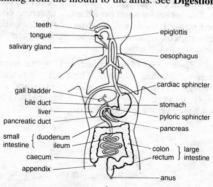

teeth
tongue
salivary gland
epiglottis
oesophagus
cardiac sphincter
gall bladder
bile duct
liver
pancreatic duct
small intestine { duodenum ileum
caecum
appendix
stomach
pyloric sphincter
pancreas
colon
rectum } large intestine
anus

Alkali A **base** which is soluble in **water**. They are usually *metal hydroxides*, e.g. sodium hydroxide. **Ammonia** solution is also an alkali. Alkalis are corrosive and should be handled with care. Typical properties of alkalis are:

− they turn red **litmus** blue
− they neutralize **acids**
− they have a **pH** of more than 7
− they react with acids to produce a **salt** and **water** only.

Alloy A **mixture** which is made up of two or more **metals** or which contains metals and **nonmetals**.

The properties of an alloy are different from the sum of the properties of the substances it contains. Alloys are used much more than pure metals. This is because an alloy can be made with a particular set of properties by bringing two or more **elements** together in the right proportions. **Aluminium** is a soft metal, but mixing it with a small amount of **copper** produces the alloy *duralumin* which is strong enough to be used in aircraft frames. Here are some other examples of alloys in common use.

Alloy	Elements it contains
brass	copper and zinc
bronze	copper and tin
pewter	**lead**, tin and small amounts of antimony
solder	lead and tin
steel	**iron** and **carbon**

Alpha particle (α) This consists of two **neutrons** and two **protons** bonded together and thus it carries a positive **charge**. It contains the same particles as the

nucleus of a helium **atom**. Many **radioactive** nuclei give off alpha particles when they decay. An example is the common **isotope** of uranium,

$$^{238}_{92}U \rightarrow {}^{234}_{90}Th + {}^{4}_{2}\alpha + energy$$

The flow of alpha particles from a radioactive source is called alpha **radiation**. The energy transferred appears as the **kinetic energy** of the alpha particles. Subsequently, alpha particles collide with other particles and slow down: there is energy transfer to the substance giving a small temperature rise. As alpha radiation loses energy in matter, it causes ions to appear, hence it is an ionizing radiation.

Matter slows down alpha radiation very quickly. It has a range of only a few cms in air and cannot pass through even thin card.

Alternating current (AC) A **current** which flows alternately in one direction, then in the opposite direction, around a circuit. The following are examples of alternating current.

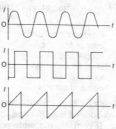

The **frequency** of the alternating current mains supply in the United Kingdom is 50 Hz (cycles per second). See **Rectifier, Direct current**.

Aluminium The most abundant metal in the earth's crust. Aluminium has **valency** 3 and is very reactive. It is obtained by the **electrolysis** of a molten mixture of the aluminium ore bauxite, dissolved in another mineral called cryolite, using graphite **electrodes**.

Aluminium and its **alloys** are resistant to corrosion due to a protective layer of oxide which forms on the surface of the metal. (This protective layer can be thickened by a process called *anodizing*). Aluminium and its alloys are widely used in the home and in industry.

Amino acids *Organic compounds* which are sub-units of **proteins**. About seventy different amino acids are known but only twenty to twenty-four are found in living organisms.

Amino Acid Structure

R   variable group (depending on amino acid)
|
NH_2—C—COOH
|
Amino   H   Acid
group   group

Amino acids are bonded together in chains known as *peptides*. The link between adjacent amino acids is called a peptide bond or link. See diagram over page.

When many amino acids are joined together in this way the whole complex is called a *polypeptide* and this is the basis of **protein** structure.

Ammonia A colourless **gas** with an unpleasant odour. It is very soluble in water and gives an alkaline (see **Alkali**) solution which is sometimes called ammonium hydroxide. Ammonia is a **covalent compound**. It has a characteristic reaction with the gas **hydrogen chloride**, producing dense white fumes of ammonium chloride.

$$NH_3(g) + HCl(g) \rightarrow NH_4Cl(s)$$

Most ammonia is made by the *Haber Process*. *Nitrogen* and **hydrogen** in the ratio of 1:3 react at 500 °C and 20 MPa pressure over an iron **catalyst**.

$$N_2(g) + 3H_2(g) \rightleftharpoons 2 NH_3(g)$$

Ammonia is produced by **bacteria** found on the roots of *leguminous* plants like peas and beans. Also, when **proteins** decompose, ammonia is released. Both of these are important sources of plant nutrients (see **Nitrogen cycle**). The following diagram shows how important ammonia is to our lives.

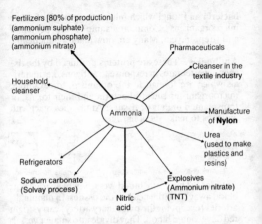

Fertilizers [80% of production]
(ammonium sulphate)
(ammonium phosphate)
(ammonium nitrate)

Pharmaceuticals

Cleanser in the textile industry

Household cleanser

Ammonia

Manufacture of **Nylon**

Urea
(used to make plastics and resins)

Refrigerators

Sodium carbonate
(Solvay process)

Explosives
(Ammonium nitrate)
(TNT)

Nitric acid

THE USES OF AMMONIA

Anhydrous This term means containing no **water** and is usually used to describe **salts** with no *water of crystallization*. For example:

	Anhydrous salt	Hydrated salt
Copper (II) sulphate	$CuSO_4$	$CuSO_4.5H_2O$
Sodium carbonate	Na_2CO_3	$Na_2CO_3.10H_2O$

The term may also be used to describe **liquids** which are perfectly dry and contain no water, e.g. anhydrous ether.

Antibiotics These are substances formed by certain

bacteria and fungi which inhibit the growth of other microorganisms. Common examples are penicillin and streptomycin. Many are now made synthetically.

Antibodies These are **proteins** produced by the **tissues** of vertebrates in response to antigens, i.e. materials which are foreign to the organism, for example, microorganisms such as **bacteria** and their toxins, or transplanted **organs** or **tissues**. Antibodies react with antigens to make them harmless.

antigens   antibodies   antigens neutralized

Arteries These are **blood** vessels which transport blood away from the heart to the **tissues**. In mammals, arteries (except for the pulmonary artery) carry bright red oxygenated blood. They divide into smaller vessels called *arterioles* which themselves eventually subdivide into **capillaries**. Arteries have thick, elastic, muscular walls as they need to withstand the high **pressure** caused by the **heartbeat**. Compare **Vein**.

Section through an artery

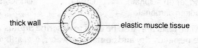

thick wall ———— elastic muscle tissue

Asexual reproduction Reproduction in which new organisms are formed from only a single parent. Asexual reproduction does not involve the production of **gametes**. The offspring are genetically identical to each other and to the parent organism. They are sometimes referred to as clones. See **Vegetative reproduction**.

Assimilation (of food) The process by which food which has already been digested is incorporated into the protoplasm of an organism. For example, in mammals, **glucose** which is not required immediately to provide **energy** in **tissue** is converted into **glycogen** by the liver and **muscle cells**. Glycogen is a short-term energy store. It can be quickly reconverted into glucose if the blood glucose level falls (see **Insulin**). Excess glucose which is not stored as glycogen is converted into **fat** and stored in fat storage cells beneath the skin. Fat is a long-term energy store. Fatty acids and glycerol are reassembled into fat within the body and stored in the same way.

Atmosphere 1. A unit of **pressure** equal to 101 325 **pascals**. Atmospheric pressure is the result of the **weight** of air pushing downwards. It varies from place to place and time to time but is always about 1 atmosphere near sea level. Atmospheric pressure decreases as height above the ground increases.

2. The **mixture** of **gases** which surrounds a planet. The atmosphere on earth is **air**. The atmosphere on earth is often divided into several regions or zones.

See diagram over page. See also **Greenhouse effect, Ozone layer**.

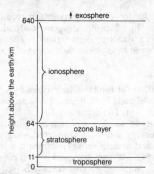

Atom The smallest particle of an **element** that can exist. Atoms are the building blocks of which everything is made. They are made up of even smaller *subatomic particles*.

Subatomic particle	Position in atom	Electric charge	Relative mass
proton	at the centre	positive	1
neutron	in the **nucleus**	neutral	1
electron	moving around the nucleus	negative	1/1840

The proton and electron carry equal but opposite charges. The atom as a whole is neutral, hence the number of protons always equals the number of electrons. All atoms of the same element have the same number of protons and hence the same **atomic num-**

ber, but atoms of the same element may have different numbers of neutrons. See **Isotopes**.

Atoms are the smallest part of an element that can take part in a chemical reaction.

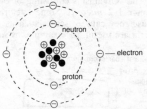

An atom of carbon

Atomic Number (Z) The number of **protons** in the **nucleus** of an **atom**. All atoms of the same **element** have the same atomic number. An atom contains the same number of **electrons** as protons, so the number of electrons also equals the atomic number.

Background radiation Radioactive matter exists in almost all of the rocks of the earth's crust. As it **decays** it gives off **radiation**. Cosmic radiation is also continually hitting the earth from space. As a result of the above two processes a radiation detector, such as a *Geiger-Muller tube*, gives a count of about one radiation a second. This is called background radiation since it is always present. Background radiation has to be taken into account when measuring the activity of other radioactive sources. The activities on earth which produce radiation, such as nuclear weapons,

13

nuclear power stations and the medical applications of radioactive **isotopes**, appear to have little if any effect on the amount of background radiation.

Bacteria Organisms consisting of only one cell of diameter 1-2 *microns*. Some bacteria are harmful because they cause diseases such as *tetanus*. Some bacteria are useful, for example, as sources of **antibiotics**.

Structure of a bacterium (generalized)

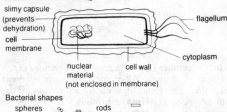

Bacterial shapes

spheres (cocci)   rods (bacilli)   spirals

Balanced diet This term is usually used in connection with humans and domestic animals. A balanced diet contains the correct nutritional components needed to remain healthy. A balanced diet for humans should contain the following:

(1) **Protein**
(2) **Carbohydrate**
(3) **Fat**
(4) **Roughage**
(5) **Vitamins**
(6) **Water**
(7) **Mineral salts**

The amounts of carbohydrate and fat should be just

enough to supply the body with its energy requirements. Excess carbohydrates and fats are undesirable as they increase the amount of fat stored in the body.

Base This is a substance which reacts with an **acid** to form a **salt** and **water** only. Bases are usually *metal oxides* or *hydroxides,* e.g. sodium hydroxide (NaOH), copper (II) oxide (CuO). Metal oxides and hydroxides which are soluble in water are known as **alkalis**. These are usually compounds of group 1 or 2 **metals**.

Battery A group of two or more single electric **cells** connected together in series. A car battery consists of six 2 V cells giving a total voltage of 12 V.

Beta particle (ß) This is a high-speed electron and thus it carries negative **charge**. It is emitted by a **nucleus** during radioactive decay or nuclear **fission**. The energy transferred appears as the **kinetic energy** of the beta particle. A flow of beta particles from a radioactive source is called beta **radiation**. Beta radiation causes ions to appear when it hits matter hence it is an ionizing radiation. Beta radiation travels further into matter than alpha radiation (see **Alpha particles**) but not as far as **gamma radiation**. It is completely absorbed by a few mms of aluminium.

Bile A green alkaline fluid which is produced in the **liver** of mammals. Bile is stored in the gall bladder and passes into the duodenum through the bile duct. In the duodenum it causes fats to be broken into tiny droplets prior to **digestion**. This process is called *emulsification* and it speeds up fat digestion.

Bimetallic strip A device made of strips of two different **metals** (hence bimetallic) fixed together. For a given **temperature** change one metal expands (see **Expansion**) more than the other. This causes the bimetallic strip to bend. The amount which it bends depends on the size of the temperature change.

Bimetallic strips are used in simple **thermometers** and **thermostats**. The following diagram shows one used in a simple fire alarm.

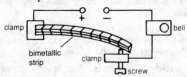

When the temperature reaches a certain value, determined by the setting of the screw, the bimetallic strip bends sufficiently to complete the **circuit** and the bell starts to ring.

Biodegradable This word describes substances which can be broken down by biological action. This action usually involves **bacteria** and **fungi**. Biodegradable waste products are usually easy to dispose of and are not a source of pollution. Nonbiodegradable materials, such as plastics, are not broken down by the action of bacteria and fungi. They are difficult to dispose of and are often a source of **pollution**.

Biosphere A collective name for the parts of the earth and the atmosphere which are inhabited by living things. Living things are found:

– in the atmosphere immediately above the earth.
– on the surface of the earth.
– within the surface of the earth.
– in lakes, rivers and oceans.
Many scientists believe that the proper functioning of the biosphere is under threat from **pollution**.

Birth (in humans) The human body is born as a result of muscular contractions of the **uterus** wall. The amniotic fluid escapes and the baby is pushed through the cervix and vagina thus leaving the mother's body.

After the baby is born the umbilical cord is cut. The **placenta** and the remaining part of the umbilical cord are then expelled from the uterus as afterbirth. The part of the umbilical cord attached to the baby will drop off within a few days. After birth the baby must use its own **lungs** for **gas exchange** and rely on its own digestive system for food. See **Fertilization**, **Pregnancy**.

Bitumen A black tarry **mixture** of high **boiling point hydrocarbons** which is left behind in the distillation of **petroleum**. It is impermeable to water and is widely used for roofing and, mixed with sand, for road surfacing.

Blast furnace A furnace which allows the continuous production of molten iron. The furnace is charged with a mixture of coke, iron ore and limestone. At the bottom of the furnace coke and **air (oxygen)** react to produce **carbon dioxide**. This reaction is **exothermic** and raises the **temperature** to 1800 °C.

$$C(s) + O_2(g) \rightarrow CO_2(g)$$

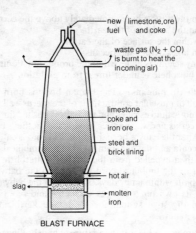

new fuel (limestone, ore and coke)

waste gas (N_2 + CO) is burnt to heat the incoming air)

limestone coke and iron ore

steel and brick lining

hot air

slag ←

→ molten iron

BLAST FURNACE

More coke then reacts with the carbon dioxide to produce carbon monoxide.

$$C(s) + CO_2(g) \rightarrow 2CO(g)$$

The carbon monoxide reduces the iron oxides to iron. For example:

$$Fe_2O_3(s) + 3CO(g) \rightarrow 3CO_2(g) + 2Fe(l)$$

The molten iron flows to the bottom of the furnace where it is tapped off.

The limestone in the charge is decomposed by the heat, producing calcium oxide and carbon dioxide.

The calcium oxide reacts with impurities from the iron ore (mainly silica SiO_2) and forms a molten slag.

$$CaO(s) + SiO_2(s) \rightarrow CaSiO_3(l)$$

The slag also sinks to the bottom of the furnace where it floats on the molten iron.

The iron produced by a blast furnace contains about 3% carbon and is very brittle. In **steel**-making the amount of carbon and other impurities is reduced by blowing oxygen into the molten iron.

Blood A fluid **tissue** which is found in many animals. The principal function of blood is to transport substances from one part of the body to another. The blood of mammals consists of a watery solution called **plasma** in which there are three types of **cells:** *platelets*, **red blood cells** and **white blood cells**.

The main functions of the blood in humans are:

(a) Transport of **oxygen** from the **lungs** to the tissues.

(b) Transport of **carbon dioxide** from the tissues to the lungs.

(c) Transport of toxic by-products to the **organs** of **excretion**.

(d) Transport of **hormones** from **endocrine glands** to target organs.

(e) Transport of digested food from the ileum to the tissues.

(f) Prevention of infection by:

   (i) **blood clotting**;

   (ii) *phagocytosis* by white blood cells;

   (iii) **antibody** production.

Blood clotting This occurs when blood is exposed to the **air** as a result of injury. The blood platelets produce an **enzyme** (thrombin) which causes the conversion of soluble plasma protein (fibrinogen) into fibrin. The fibrin forms a meshwork of fibres into which platelets become lodged. The resulting clot restricts blood loss and the entry of microorganisms.

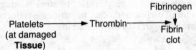

Boiling point The **temperature** at which a **liquid** boils to become a **gas**. It is not fixed but depends on atmospheric **pressure**. The higher the atmospheric pressure the higher the boiling point. Impurities in the liquid also cause the boiling point to rise. The **energy** needed to change a liquid at its boiling point into a gas is called **latent heat**.

Bond Atoms are held together in **molecules** and *giant structures* by chemical bonds. Bonds are formed between atoms and are generated by **electrons**. Some atoms lose or gain electrons forming **ions** giving rise to **ionic bonds**. Other atoms share pairs of electrons giving rise to **covalent bonds**.

Bone This is the **tissue** of which the *vertebrate skeleton* is made. It contains the **protein** *collagen* which gives *tensile strength*, and calcium phosphate which gives bone its hardness. Some bones have a hollow cavity containing *bone marrow*. New **red blood cells** are made in the bone marrow.

Brain A large mass of nerve **cells** in animals whose function is to coordinate the processes of the body. In vertebrates the brain is found at the head of the body. It is protected by the cranium and connected to the body by the spinal nerves in the **spinal cord** and by cranial nerves such as the optic nerve and the auditory nerve. The following diagram shows a generalized scheme of the vertebrate brain.

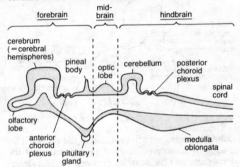

Within the human brain there are millions of nerve cells which are continually receiving and sending out *nerve impulses*. The brain is able to translate these electrical impulses in such a way that environmental **stimuli** such as **light** and **sound** are appreciated. This allows the recipient of the stimuli to respond and adapt to the **environment** as appropriate to the situation.

The brain also coordinates all body activity to ensure efficient operation. It stores information (memory) so that behaviour can be modified as the result of past experience.

Breathing (in mammals) The process of inhaling and exhaling **air** for the purpose of **gas exchange**. In mammals the gas exchange surface is situated in the **lungs**.

Lungs and associated structures

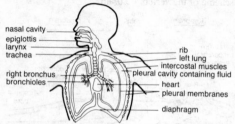

The exchange of air in the lungs (ventilation) is caused by changes in the **volume** of the thorax. This change in volume is achieved by the action of the diaphragm and the **muscles** between the ribs (intercostal muscles).

When the diaphragm contracts it depresses, and this increases the volume of the thorax. The air pressure outside the lungs (atmospheric pressure) is now greater than the air pressure inside, hence air is forced into the lungs. When the diaphragm is relaxed it reduces the volume of the thorax. Air pressure in the lungs is now greater than outside, and air is forced out of the lungs.

The action of the diaphragm is accompanied by the raising and lowering of the rib cage to accommodate the changes in the volume of the lungs. These rib cage movements are the result of contraction and relaxation of the intercostal muscles.

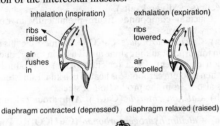

inhalation (inspiration)

ribs raised

air rushes in

diaphragm contracted (depressed)

exhalation (expiration)

ribs lowered

air expelled

diaphragm relaxed (raised)

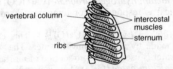

vertebral column

intercostal muscles

sternum

ribs

The rate of lung ventilation is called the *breathing rate*. The breathing rate in humans is controlled by a part of the **brain** called the medulla oblongata (see **Heartbeat**). Increased activity causes an increase in the breathing rate.

Calcium A soft grey metallic **element** from **group** 2 of the **periodic table**. It is a fairly reactive **metal** and produces a steady stream of **hydrogen** when added to cold water. Calcium compounds are found in many common rocks such as limestone and chalk ($CaCO_3$).

Calcium metal is obtained by the **electrolysis** of molten calcium chloride.

Calcium is essential to the health of plants and animals. Calcium compounds are used by plants in building cell walls and in animals for building bones, and teeth (see **Mineral salts**).

Calcium compounds figure prominently in our everyday lives.

(a) *Calcium carbonate* ($CaCO_3$). As limestone this is an important building stone and is used in the production of cement. Lime (calcium oxide) is produced when calcium carbonate is heated in a kiln. Limestone is also used in the production of steel.

(b) *Calcium chloride* ($CaCl_2$). **Anhydrous** calcium chloride is used as a drying agent.

(c) *Calcium hydrogencarbonate* ($Ca(HCO_3)_2$). This chemical is responsible for temporary **hardness of water**.

(d) *Calcium hydroxide* ($Ca(OH)_2$). This is slaked lime, made by adding lime to water. The resulting limewater is used in the production of mortar.

(e) *Calcium oxide* (CaO). This is lime. It is used in agriculture to combat excessive acidity in **soils**.

(f) *Calcium sulphate* ($CaSO_4$). This chemical is responsible for permanent hardness in water.

Camera An optical instrument which is able to produce a permanent record of a scene. Light from the scene is focused on, and absorbed by, a film. It causes chemical changes at each spot on the film. The nature of the changes depend on the colour and brightness of the light. Processing the film makes the picture perma-

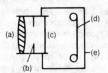

nent. The operation of the camera is as follows:

(1) The distance between the **lens** (a) and the film (d) is set so that the image is sharply focused on the film.

(2) The *iris diaphragm* (b) is put to a setting (aperture) which lets the right amount of light through.

(3) The **shutter** (c) opens, allowing the film to be exposed to the light for the right length of time.

(4) All of the parts of the camera are contained in a light-tight box which is called the camera body (e).

(5) The inside of the camera is painted matt black to prevent any light being reflected within it.

Capillaries Blood vessels formed from arterioles and eventually draining into venules and then **veins**. Capillaries form a network within vertebrate **tissue**. The capillary walls are only one **cell** thick and allow the **diffusion** of substances between the blood and the tissues via a liquid called tissue fluid (*lymph*).

Section through a capillary   Capillaries and tissues

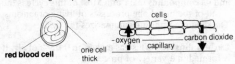

red blood cell   one cell
                      thick

Carbohydrates These are *organic compounds* which contain the elements **carbon** (C), **hydrogen** (H) and **oxygen** (O) and have the general formula CH_2O.

Carbohydrates are found either as single **sugar** units or as chains of two or more sugar units bonded together. There are three main groups of carbohydrates; *monosaccharides, disaccharides* and *polysaccharides*. Simple carbohydrates, particularly **glucose**, are the **energy** source within living **cells**. 1 g of carbohydrate contains 17.1 kJ of energy.

Long-chain carbohydrates form some structural parts of cells, for example, **cellulose** in plant cell walls. They also act as food reserves, for example, **glycogen** in animals and **starch** in plants.

Carbon A nonmetallic **element** which is in **group** 4 of the **periodic table**. It is found in nature in two different forms or *allotropes*; diamond and graphite. All living tissue contains carbon compounds *(organic compounds)*, e.g. **carbohydrates, proteins, fats,** and life would not be possible without them.

Carbon burns in air or oxygen to produce heat and light **energy**. Coke, **coal** and charcoal are all impure forms of carbon. They are often used as **fuels**.

Carbon cycle The circulation of the **element carbon** and its **compounds** in nature. Most of this cycle is the result of the metabolic processes of living organisms. The diagram opposite summarizes the cycle.

Carbon dioxide This gas makes up 0.03% of **air**. It is formed in a variety of processes.

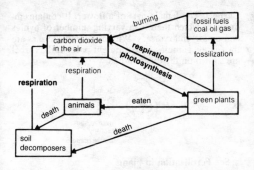

(a) By the **combustion** of **carbon** in a plentiful supply of air or **oxygen**. (Combustion in a limited supply of air will also produce the poisonous gas carbon monoxide).

(b) During the **fermentation** of **sugars**.

(c) As a waste product of **respiration**.

(d) By the action of **heat** (in most cases) or **acids** on *carbonates* in the laboratory.

Carbon dioxide plays a vital role in both **photosynthesis** and the **carbon cycle**.

The test for carbon dioxide is either that it turns lime water milky or it turns bicarb indicator yellow. The amount of carbon dioxide in the atmosphere has been steadily increasing over this century. Some scientists believe this is responsible for a small increase in the **temperature** on the surface of the earth. See **Greenhouse effect**.

Carpel The female part of a **flower**. It contains an **ovary** in which there is a varying number of **ovules** containing embryo sacs. Within the embryo sacs are the female **gametes**. The different parts of the carpel are shown on the diagram below.

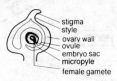

See **Fertilization in plants**.

Catalyst A substance which alters the rate of a chemical **reaction**, either speeding it up or slowing it down. The catalyst remains unused at the end of the reaction. The process is called *catalysis*. The *transition metals* or their **compounds** are often useful catalysts. Some examples are shown below.

Catalyst	*Process*
iron	hydrogen + nitrogen → **ammonia** (Haber process)
vanadium (V) oxide	sulphur dioxide + oxygen → sulphur trioxide (contact process) for making **sulphuric acid**
platinum/rhodium **alloy**	ammonia → nitric acid

Almost all of the chemical reactions which go on inside animals and plants are controlled by catalysts.

These organic catalysts are called **enzymes**.

Cell   1. A unit of cytoplasm controlled by a single **nucleus** and surrounded by a *selectively permeable membrane*. Cells are the basic units of which most living things are made. The structures of plant cells and animal cells show some similarities and some differences.

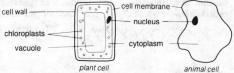

plant cell   *animal cell*

2. A device which provides a transfer to electrical **energy**. For example, a dry cell (commonly but incorrectly called a **battery**) such as might be used in a torch, gives electricity as the chemical potential energy is transferred to electrical energy.

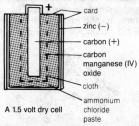

A 1.5 volt dry cell

A *photoelectric* cell provides transfer from light to

29

electric energy. Some calculators and watches are fitted with this type of cell.

Cellulose A **carbohydrate** which forms the framework and gives strength to plant **cell** walls. Humans are unable to digest cellulose; however, it has an important role as **roughage**. In mammals which are herbivores, populations of **bacteria** are present in the caecum and appendix. They produce an **enzyme** called *cellulase* which digests cellulose into **glucose**.

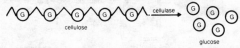

cellulose

cellulase

glucose

Chain reaction A situation where one event causes a second which, in turn, causes a third and so on. An important chain reaction is the one which occurs in a **nuclear power** station. When the **nucleus** of an **atom** of **uranium**-235 absorbs a **neutron** it undergoes nuclear **fission**. This produces two smaller nuclei, **energy** and three more neutrons. The three neutrons can, in turn, hit the nuclei of three more atoms of uranium-235 producing nine neutrons and so on.

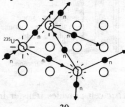

Charge This is the property of some subatomic particles. Two types of charge exist and they have the names *positive* (as on the **proton**) and *negative* (as on the **electron**).

Opposite charges attract each other while like charges repel each other. Normal **atoms** have equal numbers of protons and electrons; they are neutral and have no net charge. When electrons are gained or lost by atoms charged **ions** are produced.

Charge in motion is called **current** and involves **energy** transfer. The unit of charge is the *coulomb*, C. This is the total charge carried by about 1.6×10^{19} electrons — it is the charge transferred by one ampere in one second.

Chemical change This describes a change in which one or more chemical substances are changed into different substances by the breaking and making of chemical bonds between the atoms. Chemical change is usually accompanied by the giving out or taking in of **heat energy**. See **Exothermic reaction, Endothermic reaction**. Compare **Physical change**.

Chlorine (Cl_2) Chlorine is the second of the **halogen (group 7) elements**. It is a green **gas** at room **temperature** and is very reactive. It is very unpleasant if breathed in: it attacks the lungs and throat and produces a choking effect. Chlorine has been used as a chemical weapon.

Chlorine occurs naturally as chlorides: sodium chloride is abundant in sea water. It is extracted by the **electrolysis** of sodium chloride solution (during the industrial production of sodium hydroxide). Chlorine is used widely in a range of applications:

(a) The manufacture of chemicals such as **polymers** (PVC), **pesticides** (DDT), disinfectants (TCP) and solvents (for dry cleaning).

(b) Chlorine is added to drinking water and to water in swimming pools in order to kill dangerous **bacteria**. The process is called chlorination.

Chlorine is a vigorous oxidizing agent which readily reacts with most elements. It is prepared in the laboratory by the **oxidation** of concentrated **hydrochloric acid**.

$$\text{concentrated hydrochloric acid} + \text{potassium manganate (VII)} \longrightarrow \text{chlorine}$$

Metal chlorides. These are usually **ionic compounds**, e.g. sodium chloride (NaCl), barium chloride ($BaCl_2$) (see **salts**). They react with concentrated **acids** to produce hydrogen chloride gas.

$$H_2SO_4(l) + NaCl(s) \rightarrow NaHSO_4(s) + HCl(g)$$

| concentrated sulphuric acid | sodium chloride | sodium hydrogen-carbonate | hydrogen chloride |

Nonmetal chlorides. These are **covalent compounds**. They are usually either low **boiling point liquids** such as tetrachloromethane (CCl_4) or gases such as hydrogen chloride.

Chlorophyll A green pigment found in the **chloroplasts** of plant **cells**. This pigment absorbs the light energy needed for **photosynthesis**.

Chloroplasts Structures found in the cytoplasm of green plant **cells**. Chloroplasts contain the green pigment **chlorophyll** and **photosynthesis** occurs in them.

chloroplasts in a palisade mesophyll cell

chloroplasts in the alga *Spirogyra*

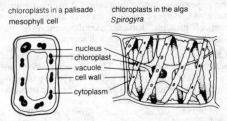

nucleus
chloroplast
vacuole
cell wall
cytoplasm

Chromatography An important technique for separating **mixtures** of solutes in a solvent. The following is a simple method of separating the different dyes in a sample of ink.

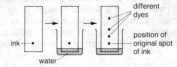

different dyes

position of original spot of ink

ink

water

(a) Place a small spot of ink onto a piece of filter paper.
(b) Stand the filter paper in a beaker containing a small amount of water. The level of water must be below the position of the ink spot. Leave the filter paper until water has risen up to the top of it.
(c) The different dyes in the ink will be carried up the

filter paper by the water at different rates hence they will be separated.

This process occurs because some substances cling to the paper more tightly than others. Chromatography has a wide range of applications including the separation of proteins, forensic samples, enzymes and viruses. It works with very small samples.

Chromosomes The hereditary material contained within the **nucleus** of a **cell**. The information they contain causes features of one generation to be passed on to the next. Each **species** has characteristic numbers and types of chromosomes.

In humans the *chromosome number* is 46. When a nucleus divides by *mitosis* this *diploid* number (46) of chromosomes is maintained in the new nuclei formed. When a nucleus divides by *meiosis* the new nuclei formed have only half of the diploid number of chromosomes (23) and they are called *haploid* nuclei. Two haploid *gametes* join to form a diploid *zygote*.

In diploid cells the chromosomes occur in similar pairs known as *homologous pairs*. Thus a human diploid cell contains 23 pairs of *homologous chromosomes*.

Chromosomes control the activity of the cell. They are made of many subunits called **genes** which contain coded information in the form of the substance **DNA**.

Circuit A number of electrical components joined by *conductors* to a source of electrical **energy**. *Circuit diagrams* show, in simple form, the order in which the components are connected. (See Appendix D for symbols used.)

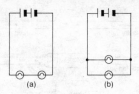

Circuits (a) and (b) show two lamps connected to two cells. In circuit (a) the bulbs are connected in *series* while in circuit (b) they are connected in *parallel*.

(c)

In circuit (c) an *ammeter* is connected in series with a lamp to measure the **current** in it and a *voltmeter* is connected in parallel to measure the **voltage** across it.

Circulatory system   Any system of vessels in animals through which fluid circulates; for example **blood** circulation and the lymphatic system.

In mammals there are two overlapping blood circulations:

(a) Circulation between the **heart** and the **lungs**.
(b) Circulation between the heart and the body.

This arrangement is called a *double circulatory system*. The heart acts as the pump for both circulations

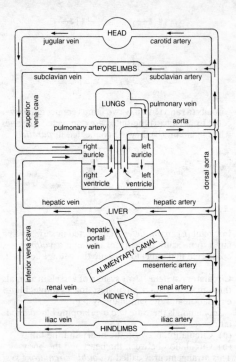

The double **circulatory system** of mammals.

and forces blood through the circulatory system always in the same direction. See diagram opposite.

Classification This is a method of arranging living organisms into groups on the basis of similarities of structure. The present system of classification was devised by Carl von Linné (Linnaeus) in the 18th century. Organisms are first sorted into large groups called *kingdoms* which are divided into smaller groups called *phyla* in animals, and *divisions* in plants. Further subdivisions occur producing subsets containing fewer and fewer organisms but which have more and more common features.

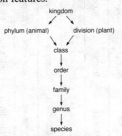

Ultimately genera (singular *genus*) are divided into groups of closely related **species**. Some examples are shown on p. 38.

Coal A **fossil fuel** formed over millions of years from fossilized plant material. It has a complex chemical structure and contains **compounds** made of car-

Classification

	human	dog	oak	meadow buttercup
kingdom	Animal	Animal	Plant	Plant
phylum/order	Chordata	Chordata	Spermatophyta	Spermatophyta
class	Mammalia	Mammalia	Angiospermae	Angiospermae
order	Primates	Carnivora	Fagales	Ranales
family	Hominidae	Canidae	Fagaceae	Ranunculaceae
genus	Homo	Canis	Quercus	Ranunculus
species	sapiens	familiaris	robur	acris

bon, hydrogen, oxygen, nitrogen and **sulphur**. Coal is used as a **fuel** to produce heat in power stations, industry and the home. The compounds containing sulphur and nitrogen produce acidic **gases** when they are burned. These gases are responsible for **acid rain**.

If coal is heated in the absence of air, coal tar is produced. Earlier in this century this was an important source of a whole range of organic chemicals such as phenol which is used in the preparation of dyes, drugs and **polymers**. Now most of these products are made from **petroleum**; however, as the world's petroleum reserves decrease, scientists are trying to find ways of converting coal into petroleum products.

About 20% of coal is used to make coke. See **Resource**.

Colour The name we give to our perception of different wavelengths of visible **light**. The colour of visible light varies with wavelength. The wavelength range of visible light is about 400—760 nm. This range contains several hundred *hues* (distinguishable colours) which merge into each other as we go from the deepest red to the deepest violet. The range is often divided into six or seven bands.

Wavelength/nm	Colour
400—420	violet
420—450	indigo
450—500	blue
500—560	green
560—600	yellow
600—650	orange
650—760	red

White light is a mixture of equal amounts of all the different hues. **Refraction** of a parallel beam of white light by a prism can produce the **spectrum** of colours

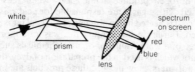

shown above. This is also possible by another method using **diffraction**. The process of separating white light into its constituent colours is called **dispersion**.

Red, blue and green are the three *primary colours*. A mixture of any two of these produces a secondary colour and a mixture of all three produces white light. Mixing coloured light is called *mixing by addition*.

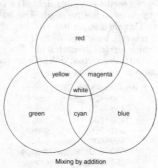

Mixing by addition

There are few surfaces which reflect all light. Most absorb some wavelengths and reflect others. An object will appear the colour of the light which it reflects. For example, when red paint is illuminated with white light the paint will absorb all colours except red. The red is reflected, making the paint look red. An object which absorbs all colours appears black.

Most paints contain several different pigments each of which will absorb particular colours. Mixing different coloured paints does not produce the same results as mixing different coloured lights.

Combustion In the process of combustion (burning), substances undergo a chemical reaction with **oxygen** (often from the air). Heat and light are usually produced during combustion and oxidation takes place.

Compound This is a pure substance which is made of atoms of two or more **elements** chemically bonded together. The **properties** of compounds are quite different from the properties of the elements from which they are made, e.g. sodium is a poisonous metal which reacts very violently with water, chlorine is a poisonous gas with a choking smell, yet sodium chloride is used in cooking and is essential to life. The atoms in a compound may be held together by either **ionic** or **covalent** bonds: e.g. methane CH_4, water H_2O, sodium chloride NaCl. See **Mixture**.

Concentration The concentration of a *solution* is a measure of how much *solute* (solid) is dissolved in the solution. It is usually expressed in terms of **mass** (e.g.

41

grams), or number of particles (e.g. **moles**) per **unit volume**. For example, the concentration of 1 dm^3 of sodium hydroxide solution containing 40 g of sodium hydroxide (relative molecular mass: NaOH = 40) could be written as:

$$40 \text{ g/dm}^3 \text{ (g/l) or 1 mol/dm}^3 \text{ (= 1 M)}$$

Conduction, electrical The process of transferring a **charge** through a medium. In nearly all cases there is also an **energy** transfer.

Substances can be classified according to the ease with which they allow a charge to pass.

(a) A *conductor* is a substance which is able to pass charge with ease.

(b) A semiconductor is fairly good at passing charge.

(c) An *insulator* will only allow charge transfer when the **voltage** applied to it is very high.

Applying a voltage to a sample involves making one end of it positive with respect to the other. If the sample contains charges which are free to move they will be drawn towards the opposite charged end.

Metals and **carbon** are the only **solids** which will allow charge to pass with ease. This is because a metal contains free **electrons** which are able to move about. The applied voltage makes the electrons accelerate. As they move they collide with each other and with the positive ions which form the bulk of the metal. These collisions cause the electrons to slow down; they introduce **resistance** to the flow of charge. There is also energy transfer which causes **temperature** rise (see **Fuse**).

The flow of charge per second is called the **current**.

It depends on the voltage and on the resistance.

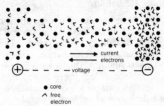

- ● core
- ʌ free electron

It is normal to give current direction as positive to negative (*conventional current*). However, when the charge carriers are electrons, the current is in fact a flow of negative charge moving from negative to positive (*actual current*).

Apart from liquid metals, other liquids will conduct if they contain positive and negative ions. These are compounds which are either molten or in solution and are called *electrolytes*. In this case the flow of charge causes chemical changes; the process is called **electrolysis**. In insulators all the electrons are either firmly attached to the atomic nuclei or involved in chemical bonding; they are not free to move.

Conduction, thermal The transfer of **energy** from particle to particle in matter. Substances differ in their ability to conduct heat. Good conductors of heat are invariably good conductors of electricity. **Metals** are good thermal conductors. They contain free moving **electrons** which carry energy rapidly from points at a high **temperature**. **Nonmetals** are very poor conduc-

tors and are used for thermal **insulation**. Compare
Convection, Radiation.

Convection The transfer of **energy** through a **fluid**
by movement within the fluid. An increase in the **temperature** of a fluid increases its **volume**. In turn, an
increase in volume causes a decrease in **density** (since
density = **mass/volume**). Thus when a fluid gets warm
it rises and displaces colder, more dense fluid. This sets
up a convection current which circulates **heat**. The
diagram below shows how heat is circulated through
the air in a room by convection currents.

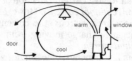

Winds and breezes are convection currents on a
huge scale in the **atmosphere**.

Cooling curve This is the general name given to a
graph of **temperature** taken as a substance cools down
over a period of time. The shape of the graph indicates
when a change of state has taken place (see **Latent
heat**). It is particularly useful for finding the **melting
point** of a **liquid** — the same temperature at which the
liquid becomes solid. It may also be used to find the
heat or temperature losses due to cooling.

Copper A *transition metal* which plays an important part in our lives. It is a vital *trace element* in our
bodies and we need between 1 — 2 mg per day.

Copper comes low down in the **electrochemical**

series and is an unreactive metal. It is sometimes found as the free (native) metal which is an indication of its lack of reactivity. It does not produce **hydrogen** when added to dilute acids.

Copper is obtained from its *sulphide* ores chalcopyrite ($CuFeS_2$) and bornite (Cu_5FeS_4), and its *carbonate* ores malachite ($CuCO_3.Cu(OH)_2$) and azurite ($2CuCO_3.Cu(OH)_2$). The final stage of purification is achieved by **electrolysis**. Impure metal is used for the *anode*, pure copper for the *cathode* and copper (II) sulphate as the **electrolyte**. Copper is widely used in plumbing and for electrical wiring.

Corrosion When a substances corrodes it is eaten away due to chemical reactions with other substances. Good examples are:
(a) The weathering of limestone buildings by rainwater which contains acids.
(b) The rusting of iron and steel due to oxidation in the presence of **air** and moisture.

Corrosion begins on the surface and often a surface layer is formed which protects the rest of the material. In rusting this is not the case. The rusting process goes through iron and some steels until it is all corroded. See **Rust**.

Covalent bonds This type of **bond** is formed when two atoms come together and share **electrons**, e.g. *hydrogen* atoms have one electron each. The atoms in a molecule of hydrogen share both electrons between them.

$$\text{H}^{\cdot} + {\,}^{\times}\text{H} \rightarrow \text{H}\,{}^{\times}_{\cdot}\,\text{H}$$

Each shared pair of electrons produces a covalent bond. Covalent bonds are usually found in **compounds** which only contain nonmetallic **elements**, e.g. carbon dioxide (CO_2), ammonia (NH_3), hydrogen chloride (HCl), and elements which exist as molecules which contain two or more atoms, e.g. hydrogen (H_2), oxygen (O_2), nitrogen (N_2), chlorine (Cl_2). Covalent bonds which contain two electrons, i.e. one shared pair, are called single bonds. Many molecules contain *double* or *triple* bonds.

ethane

ethene

$$H-C\equiv C-H$$ ethyne

Covalent compounds These are **compounds** which contain **covalent bonds**. They tend to be non conductors and have low **melting** and **boiling points**.

Current (I) The flow of **charge** through an electrical conductor (see **Conduction, electrical**). The unit of current is the ampere, A; often abbreviated to amp. One amp is the current when a charge, Q, of one coulomb passes in one second, $I = Q/t$.

Currents in **circuits** are the basis of many electrical devices and **electronic** systems. In each case the effects of charge and **energy** transfer are used. Energy transfer may produce **light, temperature** change, magnetic effects, **chemical changes** and mechanical **force**.

Decay In physics this word is used to describe how the size of a measure decreases with time. In **radioactivity** it describes how the activity of a substance decreases with time.

The following graph is a common one in either of the above contexts. It shows a pattern of exponential decay. Any value of *y* halves during a certain period of time.

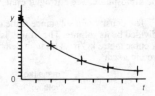

The way in which the number of active nuclei in a radioactive sample changes with time follows the above pattern. See **Half-life**.

Decomposition The breaking up of a **compound** into other compounds and elements. Heat is sometimes used to bring this about (thermal decomposition). Here are some examples:

$$2Cu(NO_3)_2(s) \rightarrow 2CuO(s) + 4NO_2(g) + O_2(g)$$

copper (II)   copper (II)   nitrogen (IV)   oxygen
nitrate   oxide   oxide

47

$$2HgO(s) \rightarrow 2Hg(l) + O_2(g)$$

mercury   mercury   oxygen
oxide

$$CaCO_3(s) \rightarrow CaO(s) + CO_2(g)$$

calcium   calcium   carbon
carbonate   oxide   dioxide

Denitrification This process occurs in **soil** and involves the conversion of nitrates into nitrogen by denitrifying bacteria. Denitrification causes a decrease in the fertility of the soil as the nitrogen is lost to the **atmosphere**. See **Nitrogen cycle**.

Density The density of a substance is the **mass** of a sample divided by its **volume**. The **SI** unit is the kilogram per cubic metre kg/m^3, although density is also often given in g/cm^3.

$$\text{density } (kg/m^3) = \frac{\text{mass } (kg)}{\text{volume } (m^3)}$$

The density of water is $1000 \ kg/m^3$ (this is equal to 1 g/cm^3).

As **solid** and **liquid** volumes change little with changes of **temperature** and **pressure** their density is fairly constant. By contrast, the density of a gas varies widely with temperature and pressure.

Dental formula A formula which describes the number and type of **teeth** in the jaws of an adult mammal with a complete set of teeth. It is expressed by writing the number of teeth in the upper jaw on one side of the mouth over those in the lower jaw on the same side. The following diagram shows how the dental formula of a human being is derived.

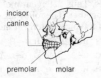

dental formula – incisor $\frac{2}{2}$ canine $\frac{1}{1}$ premolar $\frac{2}{2}$ molar $\frac{3}{3}$

total number of teeth = 2 × dental formula
= 2 × 16 = 32

It is not necessary to give a formula for the whole jaw as the teeth of mammals are always symmetrical about the centre of the jaw. In humans the upper and lower jaws are also symmetrical, however this is often not the case for other mammals.

Detergent A cleaning agent used in washing articles such as clothes and dishes. During washing a detergent acts in two ways:
(a) It reduces the surface tension of the water thus allowing the water to wet things more thoroughly.
(b) It brings together the water and (normally insoluble) fat, oil or grease to form an emulsion.
If the presence of a detergent is accompanied by rapid movement, such as the washing action of a washing machine, then particles of dirt and grime are removed and the article is cleaned.

Detergents are able to form an emulsion because one part of their **molecule** is ionic and another part

49

(the **hydrocarbon** chain) is covalent. The ionic part is attracted to water while the covalent part is attracted to oil molecules. Thus detergent molecules hold the water and oil together. The following diagram shows the structure of a typical detergent molecule.

$$H-\overset{\overset{\displaystyle H}{|}}{\underset{\underset{\displaystyle H}{|}}{C}}-\overset{\overset{\displaystyle H}{|}}{\underset{\underset{\displaystyle H}{|}}{C}}-\overset{\overset{\displaystyle H}{|}}{\underset{\underset{\displaystyle H}{|}}{C}}-\overset{\overset{\displaystyle H}{|}}{\underset{\underset{\displaystyle H}{|}}{C}}-\overset{\overset{\displaystyle H}{|}}{\underset{\underset{\displaystyle H}{|}}{C}}-\overset{\overset{\displaystyle H}{|}}{\underset{\underset{\displaystyle H}{|}}{C}}-\overset{\overset{\displaystyle H}{|}}{\underset{\underset{\displaystyle H}{|}}{C}}-SO_3^{\ominus}\ Na^+$$

covalent part (hydrophobic = 'water-hating')   ionic part (hydrophilic = 'water-loving')

Soaps work in a similar way to detergents; however their action is generally less vigorous. Soaps are made from natural oils and are sodium salts of long chain carboxylic acids. Detergents are made from petroleum.

Diffraction The bending of a wave round the edge of an opaque object into the region of shadow. The effect of diffraction on water waves can be seen in harbours or with a ripple tank.

All waves can diffract around suitable objects. The

effect is greatest when the size of the object is about the same as the wavelength of the wave.

Light diffracted by a narrow slit produces a set of *interference* fringes. As diffraction depends on wavelength, it can produce **dispersion**. A *diffraction grating* can be used to produce spectra.

Diffusion This is the way in which **fluid** particles spread from a source through the space available. For example, if a **gas** with a distinct odour (such as hydrogen sulphide) is released in the corner of a room it takes very little time before people all over the room will be able to smell it. Diffusion in **liquids** is not as fast as in gases, because the particles move at slower speeds and collide more often. Diffusion is part of the evidence for the **kinetic model** of matter.

Digestion The breakdown of large insoluble food particles into small soluble particles by the action of **enzymes**. This occurs prior to **absorption** (see **Absorption of food**) and **assimilation**. In many animals, including mammals, digestion and absorption take place in the **alimentary canal**. The table overleaf gives some details of the digestive enzymes found in the human alimentary canal.

Diode A diode is a two-electrode device which will allow an electric current to pass in one direction only (see diagram on p. 53).

The main use of diodes is as **rectifiers**. Older types of diode include diode valves and the point-contact diode ('cat's whisker').

Digestive enzymes in humans

location	glands	enzyme	substrate	product
mouth	salivary	amylase	starch	maltose
stomach	gastric	pepsin	protein	peptides
		rennin	milk protein	coagulated milk
duodenum	pancreas	amylase	starch	maltose
		lipase	fats	fatty acids + glycerol
		trypsin	protein + peptides	amino acids
ileum		lactase	lactose	glucose + galactose
		lipase	fats	fatty acids + glycerol
		maltase	maltose	glucose + fructose
		peptidase	peptides	amino acids

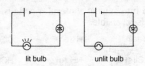

lit bulb   unlit bulb

The modern p–n junction semiconductor diode has now replaced other types for almost all applications.

Direct current (DC) A flow of **charge** in one direction only. Direct current may be constant or may vary in value. See **Rectifier, Alternating current**.

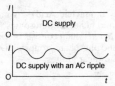

Dispersion When **radiation** bends, due to **refraction** or **diffraction**, the angle of bending depends on the wavelength. These effects can be used to separate or disperse radiation of different wavelengths. For example, *white light* consists of light rays of different wavelengths (see **Colour**). When white light passes from air into glass, the rays bend progressively, more going from the reds to the blues. The white light is dispersed.

Dissociation When a **compound** dissociates it

53

breaks up into smaller **molecules** or **ions**. This dissociation is reversible.

$$HCl(aq) \rightleftharpoons H^+(aq) + Cl^-(aq)$$
$$NH_4Cl(s) \overset{heat}{\rightleftharpoons} NH_3(g) + HCl(g)$$

dm³ This is the symbol for the cubic decimetre. The cubic decimetre is used by scientists to measure volume. In everyday life, volume is measured using another unit, the litre. One litre is exactly equal to one cubic decimetre.

$$1000 \text{ cm}^3 = 1 \text{ dm}^3 = 1 \text{ litre}$$

DNA (Deoxyribonucleic Acid) A major constituent of **genes** and hence **chromosomes**.

A *polynucleotide* chain consists of a series of alternative sugar and phosphate groups with *nitrogen bases* attached to the sugar groups.

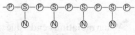

where (P) = phosphate group
(S) = sugar group
(N) = nitrogen base

DNA consists of a double polynucleotide chain twisted into a helix. The two chains are held together by bonds between nitrogen base pairs. These nitrogen bases can only link as complimentary pairs: thymine with adenine, and guanine with cytosine. The number and sequence of the base pairs in the DNA polynucleic chain represents coded information (the *genetic code*)

54

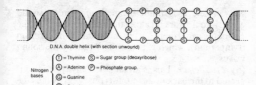

D.N.A. double helix (with section unwound)

Nitrogen bases {
T = Thymine S = Sugar group (deoxyribose)
A = Adenine P = Phosphate group.
G = Guanine
C = Cytosine
}

which allows the transfer of hereditary information from generation to generation.

Ear This **organ** is responsible for hearing and balance in vertebrates. Hearing is a sensation which is produced by vibrations or **sound waves**. These are converted into nerve impulses by the ear and transmitted to the **brain** where they are interpreted.

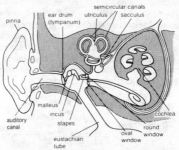

The ear is often considered in terms of three sections.

(a) *Outer ear*. The pinna is the part of the ear which

we can see. It is funnel-shaped and directs sound waves into the ear and along the auditory canal. At the end of the canal there is a thin membrane, the eardrum (tympanum), which is made to vibrate by the sound waves.

(b) *Middle ear.* This is an air-filled cavity. It is connected to the back of the mouth (pharynx) by the eustachian tube. This allows **air** into the middle ear so that the air **pressure** on each side of the eardrum is always the same. Within the middle ear there are three tiny **bones** called ossicles. The individual bones are named from their shape; malleus (hammer), incus (anvil) and stapes (stirrup). The ossicles transmit and amplify the vibrations of the eardrum to a membrane called the oval window which lies between the middle and the inner ear.

(c) *Inner ear.* This is filled with fluid and contains the cochlea and semicircular canals. The vibration of the stapes against the oval window sets up waves in the fluid of the cochlea. These waves stimulate receptor cells (hair cells) which result in impulses being sent to the **brain** via the auditory nerve. Within the brain these impulses are interpreted as sounds.

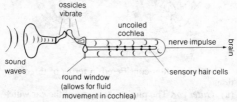

Balance is maintained by the semicircular canals in association with information received from the **eyes** and **muscles**. The canals contain fluid and receptor cells. The cells are stimulated by movement of the fluid during changes of posture. The nerve impulses initiated by these cells travel to the brain along the auditory nerve and trigger the response needed for the body to retain normal posture.

Earth   If an object is joined to earth by a conductor it will share any net **charge** it has with the earth. Because the earth is so large, this effectively means that the earthed object cannot keep charge. To earth an object it is connected to a metal plate or stake in the ground.

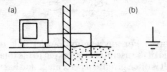

The green/yellow stripped wire in a household cable should be connected to the earth pin of the **three-pin plug**. In the event of a fault in an appliance the earth wire protects the user. Any charge flows to earth through the wire, and not through the person who touches the appliance.

Echo   This is an effect caused by **reflection. Radiation** from a source appears to come, after a short delay, from somewhere else, the image. The word is most often used in connection with **sound**.

An *echo-sounder* is a device which uses sound echoes to find depth under water. Sound from a source is reflected and the time taken for the echo to return is analysed. By knowing the speed of sound in water it is possible to relate this time to depth. **Radar** and sonar work in much the same way.

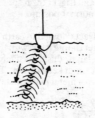

Efficiency For any **energy** transfer, this measures the ratio of the useful energy output to the total energy input. It is often given as a percentage and has no units.

$$\text{efficiency} = \frac{\text{work (energy) output}}{\text{work (energy) input}} \, (\times\, 100)$$

Of course, none of the input energy is destroyed; some is simply transferred into unwanted forms, e.g. a car engine transfers chemical energy into mechanical energy with about 25% efficiency. The majority of the energy wasted is in the form of unwanted heat.

Electrochemical series **Elements** differ in their re-activity. If a **metal** is dipped into a solution of one of its salts, for example, zinc into zinc sulphate solution, the following reaction occurs.

$$Zn(s) \rightarrow Zn^{2+}(aq) + 2e^-$$
zinc   zinc ions   2 **electrons**

The ions formed will go into the solution while the electrons cling to the remaining metal. If the metal and the solution are made part of a **circuit** the electrons will flow around the circuit. When metals are placed in order of reactivity the resulting list is called the electro-chemical series.

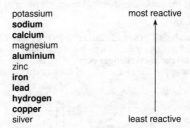

potassium   most reactive
sodium
calcium
magnesium
aluminium
zinc
iron
lead
hydrogen
copper
silver   least reactive

This series is useful because it allows us to compare reactivities and make predictions about whether certain reaction will or won't occur. An element will only displace less reactive elements from their **compounds**. Here are some examples:

— Zinc *will* remove the oxygen from copper oxide but copper *will not* remove the oxygen from zinc oxide.
— Hydrogen *will* reduce copper (II) oxide but *not* zinc oxide.

—Copper *will not* react with **acids** to release hydrogen.

Electrode A **conductor** (see **Conduction, electrical**) which dips into an **electrolyte** and allows **current** to flow to and from the electrolyte. **Copper, carbon** and **platinum** are commonly used as electrodes in **electrolysis**.

During electrolysis, the electrode connected to the positive pole of the electrical supply is called the *anode* and the electrode connected to the negative pole is called the *cathode*. Chemical reactions occur at both electrodes.

Electrolysis When a **direct** (electric) **current** is passed through a **liquid** which contains **ions** (an **electrolyte**) chemical changes occur at the two **electrodes**. For example, the electrolysis of sodium chloride solution:

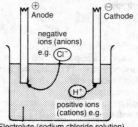

Electrolyte (sodium chloride solution)

At the anode	At the cathode
$Cl^- \rightarrow Cl + e^-$	$H^+ + e^- \rightarrow H$
$2Cl \rightarrow Cl_2$	$2H \rightarrow H_2$
chlorine is given off	**hydrogen** is given off

This electrolysis of sodium chloride is carried out on a large scale industrially to make sodium hydroxide. Electrolysis is used to extract **metals** and **nonmetals** from their **compounds**. **Aluminium**, **sodium**, and **copper** are produced in this way as is chlorine.

Electrolyte   An electrolyte is either:
(a) a molten **ionic compound**

$$NaCl(l), PbBr_2(l)$$

(b) or a solution which contains **ions**.

$$HCl(aq), NaOH(aq), CuSO_4(aq), NaCl(aq)$$

Chemical changes take place at the electrodes when a direct electric **current** is passed through an electrolyte. This process is called **electrolysis**. With molten compounds, such as sodium chloride, the changes are simple.

$$2Na^+Cl^-(l) \rightarrow 2Na(l) + Cl_2(g)$$

With ionic solutions the changes are more complicated. The products depend on:
(a) The **concentration** of the solution.
(b) The voltage which is applied.
(c) The type of **electrode** which is used.
(d) The nature of the ions present in the solution.
The following example shows what happens when

61

copper sulphate solution is electrolysed using different types of anode.

	Anode	Reaction
(a)	carbon or platinum	$4OH^-(aq) \rightarrow 2H_2O(l) + O_2(g) + 4e^-$
(b)	copper	$Cu(s) \rightarrow Cu^{2+}(aq) + 2e^-$

In process (a) oxygen is released while in process (b) the copper anode dissolves. Process (b) is used in the purification of copper.

Electromagnetic waves A range of **radiations** which differ in wavelength but have the following properties in common.
(a) They are produced by moving electric **charge**.
(b) They travel by vibrating electric and magnetic (hence electromagnetic) fields.
(c) They move through empty space at the **speed of light** (300 000 000 m/s).
(d) They tend to be absorbed by matter in which they travel more slowly.
(e) Like all waves they show **reflection, refraction,** *interference* and *diffraction* effects.
(f) They are *transverse waves* and show polarization effects.
(g) They can show particle properties.
The **spectrum** of electromagnetic waves is usually divided into a number of regions. The following table shows approximate values for the wavelengths and **frequencies** of the different regions.

Region	Wavelength/m	Frequency/Hz
gamma	$- 10^{-12}$	$10^{21} -$
X-ray	$10^{-12} - 10^{-10}$	$10^{18} - 10^{21}$
ultraviolet	$10^{-10} - 10^{-7}$	$10^{15} - 10^{18}$
visible light	$10^{-7} - 10^{-6}$	$\approx 10^{15}$
infrared	$10^{-6} - 10^{-3}$	$10^{12} - 10^{15}$
microwave	$10^{-3} - 10$	$10^{7} - 10^{12}$
radio	$10 - 10^{6}$	$10^{2} - 10^{7}$

Electron A very small subatomic particle. Electrons carry a negative **charge** and move around the nuclei of an atom. **Ions** are formed when atoms lose or gain electrons.

gaining electrons — negative *anions*, e.g. Br^-, Cl^-, O^{2-}, S^{2-}
losing electrons — positive *cations*, e.g. Al^{3+}, Fe^{2+}, H^+, Na^+

The way in which electrons are located in an atom is called the *electronic configuration*. An electric **current** is a flow of electrons moving through a **conductor**.

Element A pure substance which cannot be broken down into anything simpler by chemical means. There are 92 elements which occur naturally on the earth, and a small number of others have been made artificially in laboratories by nuclear reactions. All elements have a unique number of **protons** in their atoms.

Embryo 1. A young animal developed from a zygote as a result of repeated cell division. During **pregnancy** in mammals, the embryo develops within the female **uterus**. When the main features of the embryo have developed it is called a **foetus**.
2. A young flowering plant developed from a ferti-

lized **ovule**. In seed plants this is enclosed within a **seed** prior to **germination**.

Empirical formula The empirical formula of a **compound** shows the different atoms that are present in the molecule in their simplest whole number ratio. here are some examples:

Compound	Molecular formula	Empirical formula
Ethene	C_2H_4	CH_2
Butane	C_4H_{10}	C_2H_5
Propane	C_3H_8	C_3H_8
Ethanoic acid	$C_2H_4O_2$	CH_2O

In many compounds the empirical and molecular formulae are the same. See **Molecular formula**.

Endocrine glands (or **ductless glands**) Structures found in vertebrates and some invertebrates. They release chemicals called **hormones** directly into the bloodstream. The rate at which the hormones are secreted is often a response to changes in internal body conditions; however, it may also be a response to environmental changes.

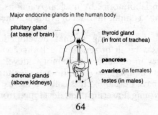

Major endocrine glands in the human body

pituitary gland
(at base of brain)

thyroid gland
(in front of trachea)

pancreas

ovaries (in females)

adrenal glands
(above kidneys)

testes (in males)

Endothermic reaction A reaction in which **heat energy** is taken in from the surroundings, so that heat has to be supplied continuously to make the reaction occur; e.g. the thermal decomposition of calcium carbonate.

$$CaCO_3(s) \rightarrow CaO(s) + CO_2(g)$$

In endothermic reactions, at about room temperature there is a drop in **temperature**; e.g. when sodium nitrate dissolves in water.

$$NaNO_3(s) \rightarrow NaNO_3(aq)$$

In endothermic reactions there is more energy in the bonds of the products than there was in the bonds of the starting materials.

Compare **Exothermic reaction**.

Energy (W) The energy of an object is the ability it has to do *work*. The unit of energy is the **joule**, J. It is often useful to speak of different 'forms of energy' (although what we really mean is energy in different contexts).

Gravitational potential energy; the work a mass, m, can do by falling a distance, h. $W = mgh$.

Electrical energy; the work an electric **current**, I, can do in time, t. $W = VIt$.

Kinetic energy; the work a mass, m, moving at speed, c, can do in coming to rest. $W = (\frac{1}{2})mc^2$.

Thermal energy; the work needed to raise a mass, m, of *specific heat capacity*, c, by θ degrees. $W = mc\theta$.

65

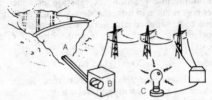

A gravitational potential → kinetic
B kinetic → rotation; rotation → electric
C electric → light + thermal

The above sketch shows a chain of energy transfers. In each case the useful energy output is less than the total energy input because some of the input energy will be transferred into unwanted forms. The ability of a device to transfer energy from one form to another is measured by its **efficiency**.

Engines   These are systems in which the chemical (potential) **energy** of a **fuel** is transferred to mechanical energy (work). The **combustion** of the fuel in **oxygen** results in a large **volume** of hot gas. As the volume of gas increases it transfers energy to the system.

Environment   A collective name for the conditions in which organisms live. Many factors contribute to the environment including:
(a) nonliving physical factors such as **temperature** and **light**.
(b) living (biotic) factors such as *predators* and *competition*.
The interaction of all of these factors determines the

conditions within *habitats* and selects the *communities* of organisms which are best suited to the prevailing conditions. See **Natural selection.**

Enzymes These are **protein** substances which act as **catalysts** within **cells.** Catalysts are substances which control the rates of chemical **reactions.** In a cell there may be hundreds of chemical reactions occurring at the same time. Each of them will require its own particular enzyme as enzymes are normally very specific in their action and will only catalyse one particular reaction. Enzymes may catalyse either:

(a) *synthesis reactions* in which complex **compounds** are produced from simple **molecules:**

(b) or *degradation reactions* in which complex molecules are broken down into simple subunits by hydrolysis.

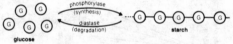

Enzymes work most efficiently within a narrow **temperature** range. Thus human enzymes work best around human body temperature (37 °C). Efficiency decreases above and below this temperature and at temperatures above 45 °C most enzymes are destroyed.

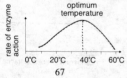

Enzymes work best at a particular **pH**. This varies for different enzymes.
(a) Salivary *amylase*, found in the saliva, works best at a neutral or slightly acid pH.
(b) *Pepsin*, found in the stomach, will only work in an acid pH.
(c) *Trypsin*, found in the intestine, works best in an alkaline pH.

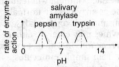

The rate of an enzyme-catalysed reaction increases as the enzyme *concentration* increases.

The rate of an enzyme-catalysed reaction also increases as the concentration of the substance on which the enzyme acts increases up to a maximum point.

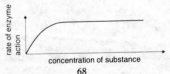

Most enzymes are named by adding the suffix -ase to the name of the substance the enzyme acts on. For example maltase is the enzyme which acts on maltose.

Equation A chemical reaction can be described by an equation. This may be a word equation:

$$\text{hydrogen} + \text{oxygen} \rightarrow \text{water}$$

or it may be a formula (or symbol) equation:

$$H_2 + O_2 \rightarrow H_2O$$

An equation is balanced if there are the same number of each type of atom on each side of the equation, e.g.

$$2H_2 + O_2 \rightarrow 2H_2O$$

is a balanced equation. There are four hydrogen atoms and two oxygen atoms on each side of it. The equation tells us that two molecules of hydrogen react with one molecule of oxygen to produce two molecules of water. Using the equation and the relative atomic masses of hydrogen (1) and oxygen (16), we can further state that 2 moles or 4 g of hydrogen reacts with 1 mole or 32 g of oxygen to produce 2 moles or 36 g of water.

$$2H_2 + O_2 \rightarrow 2H_2O$$
$$2(2) + 32 = 2(2+16) = 36$$

Sometimes *ionic equations* are used:

$$Cu^{2+} + Zn \rightarrow Cu + Zn^{2+}$$

State symbols are often added after the formula to denote which **state of matter** a substance is in, e.g.

hydrogen and oxygen **gases** react to produce liquid water.

$$2H_2 (g) + O_2 (g) \rightarrow 2H_2O(l)$$

Equilibrium In a **reversible reaction** the reaction proceeds in both directions, e.g.

$$3Fe(s) + 4H_2O(g) \rightleftharpoons Fe_3O_4(s) + 4H_2(g)$$

The reactants, iron and steam, produce the products, iron oxide and **hydrogen**. As soon as the products are made they begin to react together to reform iron and steam. Eventually, a situation is reached where the rate of the forward $\rightarrow$ reaction equals the rate of the reverse $\leftarrow$ reaction. At this point the proportion of the four substances present is constant and there appears to be no reaction taking place. The forward and reverse reactions are in dynamic equilibrium.

Ethanol One of a group of *alcohols*. It is a colourless flammable **liquid** with a **boiling point** of 78 °C. Ethanol is the alcohol in alcoholic drinks. In humans, small amounts relax the body but large amounts lead to alcohol poisoning and even death.

Alcoholic drinks are made by **fermentation**. Some of the ethanol produced by industry is also made this way, however the majority is made by the hydration of ethene.

$$C_2H_4(g) + H_2O(l) \rightarrow C_2H_5OH(l)$$
ethene   water   ethanol

$$C_2H_4(g) + H_2O(l) \rightarrow C_2H_5OH(l)$$

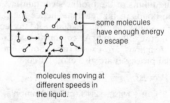

C_2H_5OH

Ethanol is used in industry as a solvent and is the main constituent of methylated spirits.

Evaporation   The change of **state** from **liquid** to **gas** which can occur at any **temperature** up to the **boiling point**. At any one time, a variable population of molecules in a liquid will have sufficient **energy** to escape into the **atmosphere**. If a liquid is left in an open container for long enough it will all evaporate.

some molecules have enough energy to escape

molecules moving at different speeds in the liquid.

In general, the lower the **boiling point** the faster the rate of evaporation, though this also depends on the **latent heat**. Liquids which are easily turned into gases are said to be volatile.

Evolution The development of complex organisms from simpler ancestors over successive generations. See **Natural selection**.

Excretion The process by which organisms get rid of the waste products of **metabolism**. The main excretory products are **water, carbon dioxide** and nitrogenous compounds such as **urea**.

In simple organisms excretion occurs through the **cell** membrane or epidermis. In higher plants excretion occurs through the leaves. Most animals have specialized excretory **organs**. For example, in humans the **lungs** excrete water and carbon dioxide and the **kidneys** excrete urea.

Exothermic reaction In an exothermic reaction heat energy is released from the reactants to the surroundings. The **bonds** of the products contain less energy than the bonds of the reactants hence the products are more stable than the reactants. Many common reactions are exothermic. All **combustion** and **neutralization** reactions are exothermic. Some important industrial processes are also exothermic, e.g.

Haber Process:

$$3H_2(g) + N_2(g) \rightleftharpoons 2NH_3(g) \text{ See \textbf{Ammonia}.}$$

Contact Process:

$$2SO_2(g) + O_2(g) \rightleftharpoons 2SO_3(g) \text{ See \textbf{Sulphuric acid}.}$$

Compare **Endothermic reaction**.

Expansion The change in size of an object or material due to **temperature** change. In terms of the **kinetic model** a temperature rise means a rise in the particles mean **energy**: i.e. the particles move more quickly and try to take up more space. This effect is greatest with **gases**. All gases expand by roughly the same amount for a given temperature change. **Solids** and **liquids** expand much less but vary greatly in the amount of expansion. See **Bimetallic strip, Water**.

Thermal expansion has many uses such as in **thermometers** and **thermostats**. It can also cause problems in structures such as clocks, roads and bridges.

Eye A *sense organ* which responds to **light**. Eyes vary in complexity from the simple structures found in invertebrates to the complex eyes of insects *(compound eye)* and vertebrates. The following diagram shows a vertical section through the human eye.

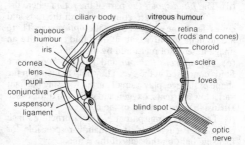

(a) The sclerotic layer is a tough protective layer which surrounds the outside of the eye. At the front of the eye it forms the transparent cornea.

(b) The choroid layer is beneath the schlerotic layer. It contains black pigmentation which prevents reflection within the eye. It is rich in blood vessels supplying the eye with food and oxygen.

(c) The retina is the inner layer of the eye. It is a layer of nerve **cells** which are sensitive to light. There are two types of cell named from their shape.

(i) *Rods* are very sensitive to low light intensity. The eyes of nocturnal animals have high concentrations of this type of cell.

(ii) *Cones* are sensitive to bright light and some are stimulated by light of different wavelengths hence they are responsible for colour vision. The *fovea* is a small area of the retina at the back of the eye which has many cones but no rods. It gives the greatest degree of detail and colour.

(d) The *blind spot* is the part of the retina where the nerve fibres, which are connected with the rods and cones, leave the eye and enter the *optic nerve.* There are no light-sensitive cells at the blind spot hence an image here is not registered by the brain.

(e) The *aqueous humour* and *vitreous humour* are fluids contained in the eye. They help to maintain the shape of the eye, play a small part in focusing the light and allow nutrients, oxygen and wastes to diffuse (see **Diffusion**) into and out of the eye cells.

(f) The *lens* is a transparent *biconvex* structure which can change curvature. It is mainly responsible for focusing light on the cells of the retina.

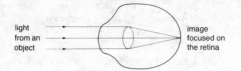

light
from an
object

image
focused on
the retina

Accommodation is the ability of the eye to focus on objects of varying distances from it. This is possible by altering the curvature of the lens.

The lens is held in place by *suspensory ligaments* which are attached to *ciliary muscles*. When the ciliary muscles are relaxed the **pressure** of the fluid in the eye keeps the suspensory ligaments taut so the lens is pulled out and its centre is thin. This is the normal relaxed state of the eye when looking at distant objects. To look at closer objects, the ciliary muscles contract. The suspensory ligaments become slack and the lens goes fatter at its centre.

The amount of light entering the eye is controlled by the *iris*. This is the coloured part of the eye and contains muscles. The hole in the centre of the iris through which light enters is called the *pupil*. In poor light the pupils are wide open (dilated) to allow the maximum amount of light into the eye. As the light gets brighter the pupil becomes smaller (contracts). This protects the retina from possible damage. This mechanism is an example of a **reflex action**. Because the pupil is small the light rays enter the eye in such a way that they produce an image which is upside down (inverted) on the retina. This is corrected by the brain. See diagram over page.

75

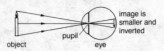

object   pupil   eye   image is smaller and inverted

Long sight and short sight are two common eye defects which can be corrected by **lenses**.

(a) A person with long sight cannot clearly see close objects because the light focuses behind the retina. This can be corrected by wearing converging (convex) lenses.

converging lens

near object

(b) A person with short sight cannot clearly see distant objects because the light focuses in front of the retina. This can be corrected by wearing diverging (concave) lenses.

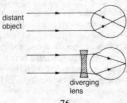

distant object

diverging lens

Fats (or lipids) *Organic compounds* which contain the **elements carbon, hydrogen** and **oxygen**. Fats are made up of three *fatty acid* **molecules** (which may be the same or different) bonded to one glycerol molecule. Fat deposits under the skin act as a long-term energy store. 1 g of fat contains 39 kJ of energy. These deposits also provide **heat insulation**.

Fat is also an important constituent in the **cell** membrane. Its insolubility in water is utilized in the waterproofing systems of many organisms.

Fermentation The process by which certain organisms such as **bacteria** and **yeasts** degrade organic **compounds** in the absence of **oxygen** in order to release **energy**. Fermentation is a form of anaerobic **respiration**. The following reaction shows fermentation by yeast (the basis of brewing and bread-making).

$$\text{glucose} \xrightarrow{\text{yeast}} \text{ethanol} + \text{carbon dioxide} + \text{energy}$$
$$C_6H_{12}O_6 \qquad\qquad 2C_2H_5OH \qquad 2CO_2$$

Fertilization The fusing of haploid **gametes** during **sexual reproduction**. It results in the formation of a single cell called the *zygote* which contains the diploid number of **chromosomes**.

(a) *External reproduction.* This occurs when the gametes are passed out of the parents and fertilized, and development takes place outside the parents. External fertilization is common in aquatic organisms, such as fish and frogs, where movement of water helps the gametes to meet.

(b) *Internal fertilization*. This is particularly associated with terrestrial animals such as insects, birds and mammals. It involves the union of the gametes within the female's body. There are several advantages to internal fertilization:

(i) the sperms are not exposed to unfavourable dry conditions.
(ii) the chances of fertilization occurring are increased.
(iii) the fertilized **ovum** is protected within a shell (birds) or within the female body (mammals).

(c) *Fertilization in humans*. During the process of copulation the penis is inserted into the vagina and **sperm cells** (produced in the **testes**) are passed out of the penis. The sperms move through the **uterus** and into the oviducts. If an ovum is present in an oviduct, fertilization can occur there. The fertilized ovum (zygote) continues moving down the oviduct towards the uterus dividing repeatedly as it does so. When it arrives at the uterus the zygote, which is now a ball of cells, becomes embedded in the prepared wall of the uterus. This process is called implantation. All further

kidney
ureter
bladder
prostate gland
Cowper's gland
penis
epididymis

spermatic cord
sperm duct
urethra
testis

kidney
ureter
bladder (displaced to the right)
urethra

oviduct
ovary
uterus or womb
cervix
vagina
vulva

development of the **embryo** occurs in the uterus. See **Pregnancy, Birth**.

(d) *Fertilization in plants.* During **pollination, pollen** grains are deposited on the stigmas of flowering plants. The pollen grains absorb nutrients and pollen tubes grow down through the style to the ovules in the **ovary**. A pollen grain enters an ovule through a small hole called a micropyle. The bottom of each pollen tube breaks down and the male gamete enters the ovule and fuses with the female gamete. After fertilization the ovule, which contains the plant embryo,

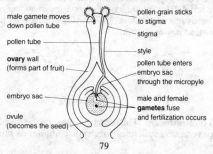

male gamete moves down pollen tube
pollen tube
ovary wall (forms part of fruit)
embryo sac
ovule (becomes the seed)

pollen grain sticks to stigma
stigma
style
pollen tube enters embryo sac through the micropyle
male and female **gametes** fuse and fertilization occurs

develops into a **seed** and the ovary develops into a **fruit**. See **Flower, Pollen, Pollination**.

Fertilizer A substance which is added to soil to increase the quantity or quality of plant growth. When crops are harvested the natural recycling of soil **mineral salts** is disturbed; i.e. mineral salts absorbed by the crops are not returned to the soil. This is called *soil depletion* and may eventually render the soil infertile. Fertilizers replenish the soil. They are often described in terms of their nitrogen, phosphorus and potassium (NPK) content. Fertilizers may be in the form of organic fertilizers, such as sewage, or inorganic fertilizers, such as ammonium sulphate. Organic fertilizers often contain trace elements: these have to be added to inorganic fertilizers.

Fission The large unstable nucleus of an atom splits into two smaller stable nuclei of similar size together with other smaller particles such as neutrons. The total mass of the products is less than that of the starting material. The difference in mass appears as energy. The fission of **uranium**-235 provides the energy in a **nuclear power** station.

Fixing nitrogen This describes any process which converts atmospheric nitrogen into **compounds** which

80

are useful as **fertilizers**. Two important processes for fixing nitrogen are:

(a) The Haber process which converts atmospheric nitrogen into **ammonia**.

(b) The action of bacteria found on the roots of leguminous plants such as beans and peas.

See **Denitrification, Nitrogen cycle**.

Flower This is the **organ** of **sexual reproduction** in flowering plants. The male part of the flower is called the **stamen** and the female part the **carpel**. Sexual reproduction relies on the transfer of **pollen** from the stamen of one flower to the carpel of the same or a different flower. See **Pollination**.

Fluid This word describes a **liquid** or a **gas**. The important feature of a substance which is a fluid is that its particles can flow and are not fixed in a particular position as is the case in a **solid**. Viscosity is a measure of the reluctance of a fluid to flow. It depends upon **temperature** because the particles 'speed' depends on temperature. A viscous liquid like syrup flows better when it is warm.

Foetus The name given to a mammalian **embryo** after the development of main features. In humans this is after about three months of **pregnancy**.

Food chain A relationship involving plants and animals in which **energy** and *carbon compounds*, made by green plants via **photosynthesis**, are passed to other living organisms, i.e. plants are eaten by animals which, in turn, are eaten by other animals.

green plant	→	herbivore	→	small carnivore	→	large carnivore
(producer)	→	(primary consumer)	→	(secondary consumer)	→	(tertiary consumer)

The arrows indicate 'is eaten by'. There are many examples of food chains, e.g.

$$grass \rightarrow sheep \rightarrow man$$

Simple food chains like the one above seldom exist. More usually food chains are linked into other food chains producing a *food web*. The following diagram shows part of the food web in a fresh water pond.

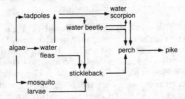

All food webs are delicately balanced. If one link should be destroyed or the numbers of a particular organism should rapidly increase or decrease, all other organisms in the web will be affected. For example, in the pond food web, if the numbers of perch started to drop due to some disease the pike population would decrease (as they would have less food) while the water scorpion population would increase (as less of them would be eaten). In turn, more water scorpions would need more food so there would be increased pressure on the populations of tadpoles and water beetles.

Force (F) A force changes an object's motion making it move more or less quickly and/or making it change direction. The unit of force is the newton, N; one newton is the force which will accelerate one kilogram by one metre per second per second. Newton's laws of motion relate forces to their effects and lead to the expression:

$$\text{force} = \text{mass} \times \text{acceleration} \ (F = m \times a)$$

Here are some effects involving forces.
(a) *Gravity.* Attraction only — why objects fall to the ground when dropped.
(b) *Magnetism.* Attraction between unlike poles (N and S) and repulsion between like poles (N and N or S and S).
(c) *Electricity.* Attraction between opposite electrical charges (+ and −) and repulsion between similar charges (+ and + or − and −).
These forces act at a distance from the object they affect. Many forces act when there is, apparently, contact between particles experiencing the force. This occurs in **friction, weight**, push and pull (compressive and tensile) forces and in twisting. A *force meter or Newton meter* is sometimes used to measure force.

Fossil fuel Substances which have been formed over millions of years from the remains of animals and plants and which are used as **fuels. Coal, petroleum** and **natural gas** are examples of fossil fuels.

All fossil fuels are composed of compounds of the **elements carbon** and **hydrogen**. When they are burned (see **Combustion**) in a plentiful supply of air they produce **carbon dioxide** and **water**. Fossil fuels also often contain small amounts of compounds containing sulphur and nitrogen which produce acidic gases when burned in air.

The world's reserves of fossil fuels are going down each year as, once used, they are not replaced by nature. Fossil fuels are nonrenewable energy **resources** (compare **Renewable energy sources**). The ever-increasing amounts of fossil fuels burned around the world each year are causing some serious problems for our **environment**. See **Greenhouse effect** and **Acid rain**.

Frequency (f) The number of cycles of a periodic motion in unit time. For example, the number of waves passing a point in one second or the number of times a pendulum swings to and fro in one minute. The **SI unit** is the hertz, Hz.

1 hertz = 1 cycle per second

The frequency, *wavelength* and **speed** of a **wave** are related to each other by the equation:

speed = frequency × wavelength ($c = f \times \lambda$)

Friction A **force** which tends to prevent motion between two surfaces. Sometimes friction works to our advantage. For example, friction prevents our feet slipping while walking or a car from skidding when driving. Unfortunately, friction is often a disadvantage. It slows down movement and wastes **energy** resulting in **heat** and wear.

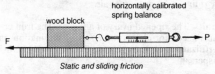

Static and sliding friction

If a block is placed on a bench and pulled gently, as shown above, at first nothing happens: the block doesn't move, thus, P = F. When P is gradually increased a maximum value is obtained before the block moves. This maximum value is the static or starting friction. Once the block starts to move, the value of P which is needed to keep it moving is called the dynamic or sliding friction. The dynamic friction is less than the static friction. If weights are put onto the block the value of the dynamic friction force increases in proportion. Even the smoothest surfaces are, in fact, rough when looked at with a microscope.

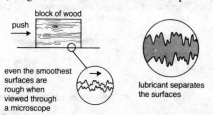

even the smoothest surfaces are rough when viewed through a microscope

lubricant separates the surfaces

Frictionless motion is obviously an advantage with transport since valuable **fuel** is used up overcoming friction. A hovercraft avoids friction with the ground by riding on a cushion of air.

Fruit The ripened **ovary** of a flower. The fruit contains **seeds** formed as the result of **pollination** and **fertilization**. The fruit protects the seed and helps in its dispersal.

Fruit and seed dispersal The methods by which most flowering plants spread their **seeds** far away from the parent plant. This is favourable to the plant because;
(a) It avoids competition for resources such as **water** and **light**.
(b) It ensures that the seeds will be spread over many different habitats, so there is a good chance that at least some of the seeds will flourish.

Most flowering plants disperse their seeds in one of the following ways.
(a) *Wind dispersal.* Air currents carry the fruits or seeds. These are usually light and are adapted to hang in the air as long as possible.

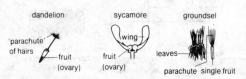

(b) *Animal dispersal.* This may happen in two different ways.

(i) Some fruits, such as burdock, are hooked and stick to the coats of animals and may be brushed off some distance from the parent plant.

(ii) Some fruits, such as the strawberry, are succulent and are eaten by animals. The succulent part of the fruit is digested but small seeds pass through the alimentary canal of the animal undamaged. Eventually they are released in the faeces, often some distance from the parent plant.

burdock   strawberry

hooks   fruits

(c) *Explosive dispersal.* Some fruits, such as sweet pea, burst open rather like an explosion. This scatters their seeds away from the parent plant. The effect is the result of unequal drying of the fruit.

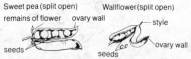

Sweet pea (split open)   Wallflower (split open)

remains of flower   ovary wall   style

seeds   ovary wall

seeds

Fuel A fuel is a substance which releases heat **energy** when it is treated in a particular way. In most fuels, energy is released by **combustion** (burning). Common fuels which produce energy by burning are paper, wood, **natural gas**, petrol and **coal**.

87

fuel + **oxygen** (often from the air) →
carbon dioxide + **water** + energy

Nuclear fuels, like **uranium** and plutonium, produce energy when their **atoms** undergo nuclear **fission**.

Fuse A fuse in an electric **circuit** melts and cuts off the supply of electricity if the **current** becomes too large. This might be due to a fault or an attempt to overload the system.

Nearly all conductors resist the flow of **charge**; the result is an **energy** transfer which causes the **temperature** of the conductor to increase. This is called the heating effect of a current.

Household fuses come in various ratings of **amperes** and are designed to melt if the current in the circuit rises above a particular value.

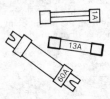

To choose the correct one for an appliance which has a **power**, *P*, in **watts**, use the equation:

$$I = P/250$$

and select a fuse which has a value slightly higher than

I. We now often use automatic switches called *circuit breakers* rather than fuses at currents higher than 13 amperes. The user can reset the switch after the fault is found and remedied.

The fuse protects the wires of the circuit and the appliance. It should be connected to the live wire. See **three-pin plug**.

Fusion (Nuclear fusion) This word is used to describe the joining of two or more light atomic **nuclei** to make a more massive one whose mass is slightly lower than the combined masses of the particles it is made from. This process involves a large transfer of **mass** to **energy**. Much of the energy from the sun and other stars comes from fusion.

$$^2_1H + {}^2_1H \rightarrow {}^3_2He + {}^1_0n + energy$$

Fusion is used in an uncontrolled way in the hydrogen atom bomb. Effective ways of controlling it have not yet been developed; however, it promises to be a cheap and clean source of power in the future.

Gamete A reproductive **cell** whose **nucleus** is formed by **meiosis** and contains only half of the normal number of **chromosomes** *(haploid)*. In humans the male gametes are **spermatazoa** and the female gametes are **ova**.

During **fertilization** male and female gametes fuse to form a *zygote*. The nucleus of the zygote contains the normal number of chromosomes *(diploid)*. Thus in humans;

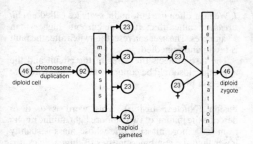

Gamma radiation (γ) High-energy (short wavelength) **electromagnetic waves** produced during the decay of certain **nuclei** (see **Radioactivity**). **Energy** leaves as a packet of radiation (sometimes called a *quantum* or *photon*), and the nucleus becomes more stable.

Gamma radiation can pass through matter very easily. It forms ions when it collides with atoms of the material and slowly loses energy. It is able to penetrate deep into the human body and may cause disorders such as cancers. Workers dealing with materials which emit gamma radiation must be protected by lead and/ or concrete shielding. *Gamma radiography* is used to examine the internal structure of metal objects such as aircraft engine components. Faults which are not visible to an exterior examination can be detected and the component replaced before the engine fails.

Gas (or vapour) The normal **state of matter** whose particles have the highest **energy**. When **heat** is supplied to a **liquid** the **atoms** or **molecules** are given **kinetic energy**. This may be enough to overcome the forces of attraction which hold the particles together in the liquid state. If this happens the liquid boils and turns to gas. Collisions of gas particles on the walls of a container exert a **pressure**. The particles move quickly, at hundreds of metres per second, in random directions and fill all the space available. The **density** of a gas is much less (e.g. 1/1000 times) than a **solid** or a liquid. Gas is a major energy **resource**.

Gas exchange In this process organisms exchange gases with the **environment** for the purpose of **metabolism**. Most organisms need a continuous supply of **oxygen** for the energy-producing reaction of **respiration**.

glucose + oxygen → **energy** + **carbon dioxide** + **water**

In addition to this green plants need carbon dioxide for **photosynthesis**.

$$\text{carbon dioxide} + \text{water} \xrightarrow{\text{light energy}} \textbf{carbohydrate} + \textbf{oxygen}$$

Both reactions use and produce gases which are exchanged between the organism and the atmosphere (land organisms) or water (aquatic organisms).

$$\text{oxygen} \underset{\text{photosynthesis}}{\overset{\text{respiration}}{\rightleftharpoons}} \text{carbon dioxide} + \text{water}$$

The gas-exchange surfaces within organisms have certain characteristics which allow the exchange to take place.

(a) A large surface area for maximum gas exchange.

(b) The surface is thin to allow **diffusion**.

(c) The surface is moist as gas exchange takes place in solution.

(d) In animals, the surface has a good **blood** supply as it is the blood which transports gases to and from the cells of the body.

In mammals gas exchange occurs across the alveoli in the **lungs**. It is the result of differences between the concentrations of oxygen and carbon dioxide in the air inside the alveoli and the deoxygenated blood in the **capillaries** around the alveoli. The differences in concentration are called *concentration gradients*. These gradients cause oxygen to diffuse from the alveoli into **red blood cells** in the capillary, and carbon dioxide to diffuse out of the blood into the alveoli. See **Breathing in mammals**.

Alveoli and associated blood vessels

deoxygenated blood from heart

oxygenated blood to heart

Gas exchange in the alveolus

oxygen

red blood cells

blood capillary

carbon dioxide

General formula This is used in organic chemistry. It is a formula which shows the relative numbers of the different atoms in terms of a variable n for all of the members of a group of compounds. The actual formula for any particular compound in the group is found by substituting a number for n. For example; the general formula for the group of compounds called alkanes is C_nH_{2n+2}. The formula for each member of the alkanes is obtained by substituting for n.

n	Name of compound	Molecular formula of compound
1	methane	CH_4
2	ethane	C_2H_6
3	propane	C_3H_8
4	butane	C_4H_{10}
etc.		

A series of compounds with the same general formula is sometimes referred to as a *homologous series*. See **Hydrocarbons**.

Generator A machine which converts **kinetic energy** to **electricity**. *Alternators* give **alternating current**; other types of generator give **direct current**. Simple forms of both AC and DC generators have a similar structure.

Genes The subunits of **chromosomes**. They consist of lengths of **DNA** and control the hereditary characteristics of organisms. A single gene consists of up to 1000 *base pairs* in a DNA **molecule**. The sequence of the base pairs represents coded information known as the *genetic code*. This code determines the structures of the different types of **proteins**, particularly **enzymes**, synthesized by the **cell**. In turn, these deter-

93

mine the structure and function of the cells and **tissues**, and ultimately the organism.

The genetic code is an arrangement of *nitrogen base pairs* in DNA. Each group of three adjacent base pairs (called triplets) is responsible within the cell for linking together **amino acids** to form protein. For example, the base triplet GTA codes for the amino acid *histidine* while GTT codes for another amino acid, *gutamine.*

The sequence and type of amino acids determine the nature of the protein. Consider two fruit flies (*Drosophila*): one has a light body colour controlled by a gene X, and the other a dark body colour controlled by a gene Y.

gene X ⟶ enzyme X ⟶ pigment X light
    synthesizes   catalyses the   body
                            formation of

gene Y ⟶ enzyme Y ⟶ pigment Y dark
    synthesizes   catalyses the   body
                            formation of

Germination The first stage in the **growth** of **spores** and **seeds**. This often follows a period of dormancy and normally requires particular environmental conditions such as the availability of water and **oxygen** and a favourable **temperature**. If these conditions are not present, spores and seeds can often remain alive for some time (often years) before germinating. In this state they are said to be *dormant.*

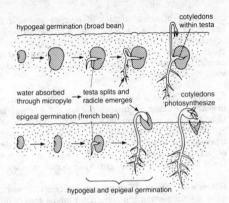

hypogeal germination (broad bean)

cotyledons within testa

water absorbed through micropyle

testa splits and radicle emerges

cotyledons photosynthesize

epigeal germination (french bean)

hypogeal and epigeal germination

Seed germination in flowering plants. There are two types of germination; hypogeal and epigeal. The difference lies in what happens to the cotyledons. In epigeal germination the cotyledons become the first leaves of the new plant while in hypogeal germination they do not. In both cases water is absorbed through the micropyle, the testa splits open and the radicle emerges.

Glucose A monosaccharide **carbohydrate** made during **photosynthesis**. This substance is an important **energy** source in both animal and plant **cells**. See **Respiration**.

Glycogen This is a polysaccharide **carbohydrate**. It

consists of branched chains of **glucose** units. It is important in animals as a short-term energy store. In vertebrates glycogen is stored in muscle and liver cells. It can be rapidly converted to glucose by the action of amylase **enzymes** if the level of glucose in the blood is too low. See **Insulin**.

Gravity This is a **force**. It is an attraction between masses. The cause of gravity is not known. The **weight** of an object depends on the value of the gravitational attraction and its **mass**. Gravity is taken to act through a point at the centre of an object where we imagine the object's mass to be concentrated. This point is sometimes called the *centre of gravity* or *centre of mass*.

Greenhouse effect **Temperatures** around the world have shown a small but significant increase over the last century. Some scientists believe that this is linked to the increasing concentration of **carbon dioxide** in the **atmosphere**. The earth receives **heat radiation** from the sun. Some of this is absorbed by the earth and re-emitted back into space at a longer wavelength. The atmosphere reflects some of the re-emitted radiation back on the earth and prevents it from going into

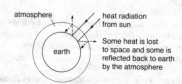

atmosphere

heat radiation from sun

earth

Some heat is lost to space and some is reflected back to earth by the atmosphere

space. The more carbon dioxide there is in the atmosphere the more heat radiation is reflected back to earth, hence the hotter the earth will become. A similar effect happens in a greenhouse; the glass reflects re-emitted heat radiation back into the greenhouse (hence the name of this effect).

The increasing concentration of carbon dioxide in the atmosphere is a result of the combustion of increasing quantities of fossil fuels and the large scale deforestation in many parts of the world (less vegetation therefore less **photosynthesis**). The greenhouse effect is a serious problem because it is believed to be responsible for changes in the climate of some areas of the world and it may result in an increase in the sea level as part of the ice caps melt.

Group In the **periodic table** the **elements** are arranged in horizontal *periods* and vertical groups.

group 1	lithium, sodium, potassium
group 2	beryllium, magnesium, calcium
group 3	boron, aluminium
group 4	carbon, silicon
group 5	nitrogen, phosphorus
group 6	oxygen, sulphur
group 7	fluorine, chlorine, bromine
group 0	helium, neon, argon

In each group the outermost electron shell contains the same number of **electrons** for each member. The number of electrons is the same as the group number.

Growth The increase in size and complexity of an

organism during its development from **embryo** to maturity. It is the result of cell division, cell elongation and cell differentiation. In plants growth originates at certain localized areas called *meristems*. In animals growth goes on all over the body.

Half-life period ($t_\frac{1}{2}$) The time it takes for a measure, whose **decay** is exponential, to fall to half of its value. Half-life is often used in connection with the number of parent radioactive **nuclei** in a sample.

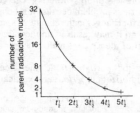

The number of parent radioactive nuclei will decrease by half every $t_\frac{1}{2}$ period.

Halogens This is a collective name for the **elements** in **group** 7 of the **periodic table**. They are all poisonous nonmetallic elements. At room temperature flourine and **chlorine** are gases, bromine is a liquid and iodine is a solid. All the halogens are oxidizing agents (see **Oxidation**); their oxidizing power and chemical reactivity decreases as we go down the group, i.e.

$$F_2 \; > \; Cl_2 \; > \; Br_2 \; > \; I_2$$

They react vigorously with **metals** and **hydrogen** forming halides:

fluorine → fluorides
chlorine → chlorides
bromine → bromides
iodine → iodides

They all contain seven **electrons** in the outer shell of the **atom** and form **ions** with a single negative charge (univalent anions) e.g. Cl^-, Br^-.

Hardness of water   Water from the mains supply is often described as being hard or soft depending on how well it lathers with soap. Soft water lathers easily with small amounts of soap whereas hard water does not. Hard water forms a scum and requires more soap than soft water to produce a lather.

Hardness in water is caused by dissolved calcium and magnesium salts. These salts are dissolved in the water when it flows over rocks such as limestone, chalk and gypsum. The metal **ions** are responsible for the hardness since they react chemically with the soap to form the scum. Soft water collects in areas of the country where the rocks are insoluble in water, e.g. the granite areas such as the Lake District, Cornwall and the Highlands of Scotland. There are two distinct types of hard water.

(a) *Temporary hard water.* This contains dissolved calcium hydrogencarbonate due to slightly acidic rain dissolving chalk and limestone.

$$H_2O(l) \quad + \quad CO_2(g) \quad \rightarrow \quad H_2CO_3(aq)$$
water   carbon dioxide   carbonic acid

$$CaCO_3(s) \quad + \quad H_2CO_3 \quad \rightarrow \quad Ca(HCO_3)_2(aq)$$
calcium   carbonic   calcium
carbonate   acid   hydrogencarbonate

It is called temporary hardness because it can be removed simply by boiling the water. This reverses the above reactions and results in insoluble calcium carbonate being deposited. In temporary hard water areas the calcium carbonate causes problems as lime scale in kettles and boilers.

(b) *Permanent hard water*. This contains dissolved calcium sulphate and magnesium sulphate. These chemicals cannot be removed by boiling but are removed by adding washing soda (sodium carbonate). This precipitates them as insoluble carbonates.

Heart A muscular pumping organ which maintains the circulation of **blood** around the body. It usually contains valves which prevent the blood from flowing backwards. In mammals the heart has four chambers; two relatively thin-walled atria (auricles) which receive blood and two thicker-walled ventricles which pump blood out. The right side of the heart deals only with deoxygenated blood and the left side only with oxygenated blood. The wall of the left ventricle is thicker and more powerful than that of the right since it pumps blood to all parts of the body whereas the right ventricle only pumps blood to the **lungs**. See **Circulatory system, Heartbeat**.

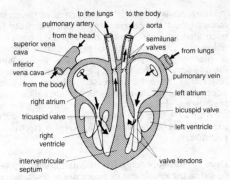

Heartbeat This is the result of alternative contraction and relaxation of the **heart**. In mammals it consists of two distinct phases:

(a) *Diastole.* The atria and ventricles relax. This allows blood to flow from the atria into the ventricles.

(b) *Systole.* The ventricles contract forcing blood into the pulmonary **artery** and the aorta. At the same time the relaxed atria fill with blood ready for the next beat.

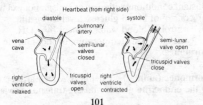

Heartbeat (from right side)

Heartbeat is initiated by a structure in the right atrium called the *pacemaker*, although the rate is controlled by a part of the **brain** called the medulla oblongata. This part of the brain monitors the activity of the body by detecting increases in the **carbon dioxide** level in the blood; the result of increased **respiration** (caused by increased activity). It is also affected by certain **hormones** such as adrenaline from the adrenal glands. The rate of human heartbeat is measured by counting the **pulse rate**. People with heart defects are sometimes fitted with electronic pacemakers.

Heat A form of **energy**. It is the internal energy possessed as the movement of **molecules** of a substance (see **Kinetic energy**). The word is also used to mean energy transfer. Thus when we heat a beaker of water with a bunsen burner there is energy transfer from the burning gas to increase the particle motion in the water. Heat transfer may be by **conduction (thermal)**, **convection** or **radiation**.

Homeostasis A general term for the maintenance of constant conditions within an organism. For example:
(a) control of **blood glucose** level by **insulin**.
(b) control of blood water content by ADH.
(c) control of body **temperature** by the **skin**, etc. See **Hormones**.

Hormones (animal) Chemicals which are secreted by the **endocrine glands** and are transported via the bloodstream to certain target **organs**. At these organs the hormones cause specific effects which are vital in regulating and coordinating the activities of the body. Hormones are sometimes called *chemical messengers*. Hormone action is usually slower than nervous stimulation. The tables on pp. 104-6 summarize the properties of some important human hormones, however there are many others.

Humus   A general term for organic material in soil consisting of decomposing plant and animal remains. It is a desirable feature of **soils** as it provides plants, and ultimately animals, with nutrients.

Hybrid   A plant or animal produced as a result of a cross between two parents who are genetically unlike each other or between two differing but related organisms.

Hydrocarbon   A **compound** which contains only **hydrogen** and **carbon**. There are three main series of hydrocarbons.

Name of series	General formula	Examples	
Alkanes	C_nH_{2n+2}	methane	CH_4
		ethane	C_2H_6
		propane	C_3H_8
Alkenes	C_nH_{2n}	ethene	C_2H_4
		propene	C_3H_6
Alkynes	C_nH_{2n-2}	ethyne	C_2H_2
		propyne	C_3H_4

Endocrine gland	Hormone	Effects
Pituitary gland	ADH (Anti-Diuretic Hormones)	Controls water reabsorption by the **kidneys**.
	TSH (Thyroid Stimulating Hormone)	Stimulates *Thyroxine* production in the thyroid gland.
	FSH (Follicle-Stimulating Hormone)	Causes ova to mature and the ovaries to produce oestrogen.
	LH (Luteinizing Hormone)	Initiates ovulation and causes the ovaries to release progesterone.
	Growth hormone	Stimulates growth in young animals. In humans, deficiency causes *dwarfism,* and excess causes *gigantism.*
Thyroid gland	*Thyroxin*	Controls rate of growth and development in young animals. In human infants, deficiency causes *cretinism.* Controls the rate of chemical activity in adults. Excess causes thinness and over-activity, and

		deficiency causes obesity and sluggishness.
Pancreas (*Islets of Langerhans*)	Insulin	Stimulates conversion of **glucose** to **glycogen** in the liver. Deficiency causes *diabetes*.
Adrenal glands	*Adrenalin*	Under conditions of *"fight, flight, or fright"* causes changes which increase the efficiency of the animal. For example, increased heart beat and breathing, diversion of blood from gut to muscles, conversion of glycogen in the liver to glucose.
Ovaries	Oestrogen	Stimulates **secondary sexual characteristics** in the female, for example, breast development. Causes the uterus wall to thicken during menstrual cycle.
	Progesterone	Prepares uterus for implantation.

Testes	Testosterone	Stimulates secondary sexual characteristics in the male, for example, facial hair.

Hydrochloric acid   A strong acid made by dissolving *hydrogen chloride* (HCl) gas in water. Hydrogen chloride gas can be made by reacting the **elements** together directly or by reacting sodium chloride with concentrated **sulphuric acid**.

$$H_2(g) + Cl_2(g) \rightarrow 2HCl(g)$$
$$NaCl(s) + H_2SO_4(l) \rightarrow NaHSO_4(s) + HCl(g)$$

The gas reacts with **ammonia** to form dense white fumes of ammonium chloride.

$$NH_3(g) + HCl(g) \rightarrow NH_4Cl(s)$$

The **atoms** in hydrogen chloride gas are held together by **covalent bonds**; however, **ions** are formed when it dissolves in a *polar* solvent such as water, hence hydrochloric acid is an ionic compound and contains hydrogen (H^+) and chloride (Cl^-) ions.

Hydrogen chloride gas is very soluble in water. The maximum **concentration** of the solution is 36% (about 11 **moles/dm³**). The acid is *monobasic* and produces **salts** called chlorides.

$$Fe(s) + 2HCl(aq) \rightarrow FeCl_2(aq) + H_2(g)$$
$$Mg(s) + 2HCl(aq) \rightarrow MgCl_2(aq) + H_2(g)$$

The acid releases **carbon dioxide** from carbonates and can be oxidized to **chlorine**.

$$Na_2CO_3(s) + 2HCl(aq) \rightarrow 2NaCl(aq) + CO_2(g) + H_2O(l)$$
$$MnO_2(s) + 4HCl(aq) \rightarrow MnCl_2(aq) + 2H_2O(l) + Cl_2(g)$$

Hydroelectricity Electrical power produced by using the **energy** of falling water, often from behind a dam.

Hydrogen (H_2) A reactive **element** which is a diatomic **gas**. An atom of the commonest isotope of hydrogen consists of a **nucleus** of one **proton** and, outside the nucleus, one orbiting **electron**. Hydrogen has two other **isotopes**, *deuterium* and *tritium*, whose nuclei also contain one and two **neutrons** respectively. Hydrogen can form **covalent** bonds by sharing electrons; e.g.

$$2H_2(g) + O_2(g) \rightarrow 2H_2O(g)$$
$$C_2H_4(g) + H_2(g) \rightarrow C_2H_6(g) \ (hydrogenation)$$

Hydrogen ions (H^+) are produced when hydrogen atoms lose their electrons. (Under the right conditions a hydrogen atom can produce hydride ions, H^-). **Acids** contain hydrogen ions. Hydrogen is a **reducing agent**. It is used to convert vegetable oils into *margarine*.

Industrially, hydrogen is made from **petroleum** by the *steam reformation of alkanes*. Large quantities of hydrogen are used in the formation of **ammonia** by the *Haber Process*. In the laboratory hydrogen is made by reacting a **metal** with an acid other than nitric acid. Zinc and dilute **hydrochloride acid** are commonly used.

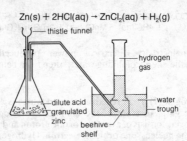

$$Zn(s) + 2HCl(aq) \rightarrow ZnCl_2(aq) + H_2(g)$$

Hydrogen is also released by the **electrolysis** of aqueous solutions which contain ions of elements above hydrogen in the **electrochemical series**, e.g. $NaCl(aq)$, $Mg(NO_3)_2(aq)$, and by the reaction of water with **group 1** and group 2 metals, e.g. sodium, calcium.

Hydrogen is highly inflammable and forms explosive mixtures with **oxygen**. It has a very low **density**, and was once used in balloons and airships, however several large-scale disasters occurred due to its highly inflammable nature.

Hydrogen ion (H⁺) Hydrogen loses a single **electron** to give a positively charged **ion** (which is, in fact, a **proton**).

$$H \rightarrow H^+ + e^-$$

This small particle is very reactive. In aqueous solution it combines with a **molecule** of water to form an *oxonium ion*.

$$H^+ + H_2O \rightarrow H_3O^+$$

All aqueous solutions of acids contain more hydrogen ions than **hydroxide ions**. The **pH** of a solution is a measure of the **concentration** of hydrogen ions in it.

Hydroxide ion (OH⁻) This is found in all alkalis and in alkaline solutions, e.g. sodium hydroxide Na^+OH^-. It is present, to a small extent, in all aqueous solutions because of the **dissociation** of **water**.

$$H_2O \rightarrow H^+ + OH^-$$
water   **hydrogen ion**   hydroxide ion

Solutions which contain more hydroxide ions than hydrogen ions are described as alkaline and they have a **pH** greater than 7.

Indicator A substance which is different colours in acidic and alkaline conditions. It can be used to show whether a solution is an **acid** of an **alkali**.

Indicator	Colour in acid	Neutral	Colour in alkali
litmus	red		blue
phenophthalein	colourless		red
universal	red orange yellow	green	blue purple

Infrared A region of the **spectrum** of **electromagnetic waves**. The approximate wavelength range is $10^{-6} - 10^{-3}$, and the approximate frequency range $10^{12} - 10^{15}$ Hz.

Infrared radiation is thermal or heat radiation. It is detectable by blackened thermometers, skin nerve cells, thermopiles and photographic films. All matter radiates infrared at all times.

Insulation   A technique to reduce the **energy** transfer from one place to another. Thus *thermal insulation;*
(a) reduces **heat** energy losses from a high temperature area such as an oven;
(b) or reduces heat energy gains to a low temperature area such as a refrigerator.

In cold climates people use many methods of insulation to reduce heat loss from their homes. These may include *double glazing, cavity wall insulation* and *loft insulation,* and make use of such materials as *fibreglass, polystyrene* and *mineral wool.* Electrical insulation often involves coating conductors with nonmetals, such as plastics, to prevent charge (and thus energy) being transferred anywhere other than along the conductor.

Metals are good conductors of both thermal and electrical energy. Their value in insulation is limited to use as reflectors of radiation.

Insulin   A vertebrate **hormone** which is secreted by the Islets of Langerhans in the **pancreas**.

Insulin regulates the conversion of **glucose** into **glycogen** which occurs in the **liver**. If the concentration of glucose in the **blood** is high, the rate of secretion of insulin is high. Thus glucose is rapidly converted into glycogen for storage in the liver. Conversely, if the concentration of glucose in the blood is low, less insulin is secreted, and less glucose is converted into glycogen. This is an example of the feedback regulation associated with many hormones. A person suffering from *diabetes* is unable to produce enough insulin to control the delicate glucose level balance in their body.

Intestine This is the region of the **alimentary canal** between the stomach and the anus or cloaca. In vertebrates, most **digestion** and absorption of food occurs in the intestine and it is usually differentiated into the small intestine and the large intestine.

Ion An atom which has become electrically charged by gaining or losing electrons. *Cations* are positively charged, e.g. Na^+, and *anions* are negatively charged, e.g. O^{2-}. Atoms tend to lose or gain electrons to produce an ion with the same *electronic configuration* as a **noble gas**. Groups of atoms (radicals) may also form ions.

sulphate SO_4^{2-} nitrate NO_3^-
hydroxide OH^- ammonium NH_4^+

Ionic bonds These are chemical bonds which occur because of *electrostatic* attractive forces between positively and negatively charged **ions**. Ionic bonds are often present in **compounds** of **nonmetals** from groups 6 and 7 and **metals**, e.g. Na^+Cl^-, $Mg^{2+}(Br^-)_2$ and in compounds containing *radicals* such as sulphate and nitrate, e.g. $Cu^{2+}SO_4^{2-}$, $K^+NO_3^-$.

Ionic compounds   These are **compounds** which contain **ionic bonds**. They tend to have high **melting** and **boiling points** and are **conductors** of electricity when molten.

Iron   The most widely used metallic **element**, although it is usually encountered as **steel alloys**. In this form it is used for building girders, car bodies, cans, tools and many other items of everyday life.

Iron is extracted from ores such as haematite (Fe_2O_3) and magnetite (Fe_3O_4) by reduction with carbon monoxide in a **blast furnace**. The iron produced in a blast furnace is brittle and is made into a range of different steels with varying properties.

A major problem in using iron and many steels is **corrosion (rust)**. The iron oxidizes in moist air to produce a soft crumbly oxide.

$$iron \xrightarrow{\text{moist air}} iron\ (II)\ oxide\ (Fe_2O_3)$$

Iron is a *transition metal* and can have **valency** 2 or 3. It reacts with dilute acids to form iron (II) compounds, e.g.

$$Fe(s) + 2HCl(aq) \rightarrow FeCl_2(aq) + H_2(g)$$

Small amounts of iron are an essential part of the human diet. It is needed by the body to make the pigment haemoglobin which gives the red colour to **red blood cells**.

Isotopes These are species of the same chemical **element**. They have the same number of **protons** but different numbers of **neutrons** in their **nuclei**. The following table shows some details of the isotopes of **carbon**.

	Number of protons	Number of neutrons
carbon-12	6	6
carbon-13	6	7
carbon-14	6	8

Joint The point in a **skeleton** where two or more **bones** meet. Some joints allow movement. There are three main types of movable joint in mammals.
(a) *Ball and socket joint.* This allows movement in several planes. In humans the shoulder and hip are ball and socket joints.

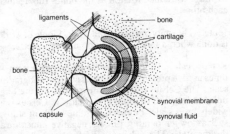

113

(b) *Hinge joint.* This allows movement in only one plane. In humans the elbow and knee are hinge joints.

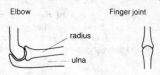

Elbow   Finger joint

radius

ulna

(c) *Gliding joint.* This type of joint occurs when two flat surfaces are able to glide over each other allowing only a small amount of movement. In humans the wrist and ankle are gliding joints.

cartilage

disc

vertebrae

The ends of bones at movable joints are covered in *cartilage* to reduce **friction** as the bones rub against each other.

Joule The unity of **energy** and work. 1 joule of work is done when a **force** of 1 newton moves 1 metre.

Kelvin scale This is a scale of **temperature** sometimes used in preference to the Celsius (or centigrade) scale. The Celsius scale is positive above the **melting point** of ice (0 °C) and negative below it. Celsius is a widely used scale; however, in some fields of study, such as low temperature physics, negative values are a disadvantage. The Kelvin (or absolute) scale does not have negative values since its zero is at *absolute zero*. To convert Celsius temperature to Kelvin temperature add 273, i.e.;

$$0° C = 273 K$$
$$100° C = 373 K$$

Kidney This organ is found in vertebrates and consists of a collection of units called *nephrons*. It is concerned with **excretion** and **osmoregulation**. In humans, the kidneys are a pair of red-brown oval structures at the back of the abdomen.

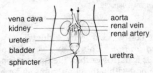

Oxygenated **blood** enters each kidney via the renal **artery** and deoxygenated **blood** leaves by the renal **vein**. A tube called the *ureter* connects each kidney with the bladder.

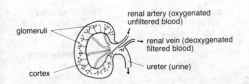

Within the kidney the renal artery divides into many small blood vessels called *arterioles* which terminate in tiny knots of blood **capillaries** called *glomeruli*. There are about one million glomeruli in a human kidney. Each is enclosed in a cup-shaped organ called a Bowman's capsule.

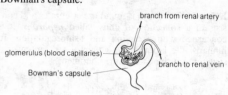

Two processes occur in the kidneys:

(a) *Ultrafiltration.* The blood vessel leaving the glomerulus is narrower than the one entering it. This puts the blood within the glomerulus under high **pressure**. Under these conditions the capillary wall acts as a selectively permeable membrane. Blood components which have only small molecules are forced through it into the Bowman's capsule while components with larger molecules remain in the blood.

Large molecules remaining in the blood	Small molecules filtered out of the blood
blood cells	glucose
plasma proteins	urea
	mineral salts
	water
	amino acids

(b) *Reabsorption.* The fluid filtered from the blood

passes from Bowman's capsule down the renal tubule. Each tubule is supplied with a network of blood vessels and as the fluid passes through it useful materials are reabsorbed into the blood. These will include all the glucose and amino acids and some mineral salts and water.

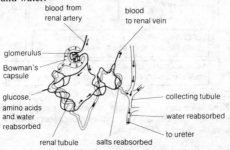

By the time the liquid reaches the end of the renal tubule it consists of a solution of urea and mineral salts. It drains into the bladder via the ureter and is eventually expelled from the body as **urine**. The amount of water which is reabsorbed is controlled by a **hormone**, called the antidiuretic hormone (ADH). It is released by the pituitary gland which monitors the amount of water in the blood. If the level of water is too low more ADH is released into the bloodstream. This results in more water being reabsorbed by the kidneys leaving the urine more concentrated. If the level of water is too high the converse occurs. See diagram overleaf.

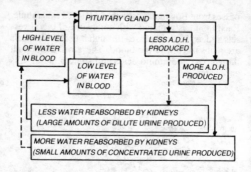

Kinetic energy An object has kinetic **energy** if it is moving. The amount depends on the object's **mass** m and its **speed** v as follows;

$$\text{kinetic energy} = \tfrac{1}{2}\,\text{mass} \times \text{speed}^2$$
$$W = \tfrac{1}{2}mv^2$$

Kinetic model This is a theory about the structure of matter which is commonly accepted by scientists. It is described by the following four statements:
(a) All matter is made of particles.
(b) The particles are always moving.
(c) The particles attract each other.
(d) The **temperature** of a substance relates to the mean **energy** of its particles.

The kinetic theory is able to explain the behaviour of matter. For example; See **Conduction, Convection,**

Diffusion, Evaporation, Expansion, States of Matter.

Latent heat This term means hidden heat. It is the **energy** involved in changes of **state**. If heat is added, at a constant rate, to a **solid**, the **temperature**/time graph has shape (a). A similar graph for a **liquid** gradually losing heat at a constant rate has the shape (b).

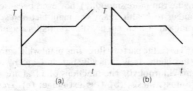

(a)   (b)

In each case, the temperature stays constant while the change of state takes place. A similar situation exists in the changes for liquid to **gas** and gas to liquid. The quantity of energy transferred to or from the particles during changes of state depends on the nature of the substance and its state.

Lead A metallic **element** in group 4 of the **periodic table**. It is a soft grey metal and has a high **density**. The main source of lead is the ore galena (PbS). Lead and its **alloys** are widely used, e.g. in the lead-acid *accumulator*, solders, and as flashing on roofs to keep out water. Lead absorbs radiation from radioactive isotopes and X-rays. It is used for screening and containing radioactive sources to protect people from the effects of radiation. Lead can be used to store highly corrosive chemicals such as **sulphuric acid**.

At one time lead was widely used for water pipes, and some of its **compounds** were the basis of paints; however, studies have shown that even very small amounts of lead absorbed into the body can damage a person's health. A lead compound, tetraethyl lead, is still widely used as an antiknock agent in petrol (makes the engine run more smoothly), however, because of the danger to health, many car manufactureres are designing cars which can use other antiknock agents.

Leaf This is the part of a flowering plant which grows from the **stem**. It is typically flat and green, and contains **chlorophyll**. The functions of a leaf are;
(a) **Photosynthesis**; (b) **Gas exchange**; (c) **Transpiration**. The following diagram shows the structure of a typical leaf.

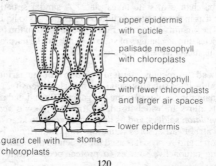

upper epidermis with cuticle

palisade mesophyll with chloroplasts

spongy mesophyll with fewer chloroplasts and larger air spaces

lower epidermis

guard cell with chloroplasts — stoma

LED This is short for light-emitting diode. It is a p–n junction **diode**, and is usually made from gallium arsenide phosphide. **Energy** is released within the LED and this is given off as light. The junction is made near to the surface so that the emitted light can be seen. No light is emitted with a reverse bias. LEDs are commonly coloured red, yellow or green. They are widely used in a variety of electronic devices.

Lens An optical device which bends transmitted

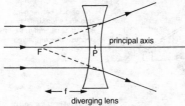

diverging lens

light by **refraction**. Here are some examples of different shapes of lenses.

converging lenses   diverging lenses

(a) *Converging (convex) lenses* bring the rays together at a point called the *focus*. They are thicker at the centre than at the edge.

(b) *Diverging (concave) lenses* spread the rays out so

they appear to be coming from one point, the focus. They are thin at the centre but get thicker towards the outside.

See **Eye**.

Lever This is a type of **machine** in which a certain **force** applied at one point gives an output force elsewhere. Levers are often grouped into three classes (1, 2 and 3). The classes differ in the positions of the input

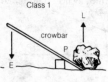

Class 1

The pivot is between the load and effort. Usually the effort is smaller than the load because it is further out.

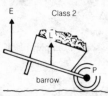

Class 2

The pivot is at the end of the lever and the load is in the middle. A small effort lifts a large load.

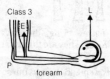

Class 3

The pivot is again at the end but the effort is in the middle. There is a *mechanical disadvantage* and a large effort is needed to lift a small load.

force (the effort applied) and the output force (the load overcome) relative to the pivot (turning point or fulcrum).

Light A region of the **spectrum** of **electromagnetic waves** which can be seen by the normal human **eye**. It has wavelengths between $400 - 760$ nm. The colour of the light is related to its wavelength. Like all waves, light can be absorbed, reflected, refracted, diffracted and shows interference effects. Like all electromagnetic waves, it travels through empty space at 300 000 km/s and this value is commonly called the **speed of light**.

Liquid A **state of matter** in which particles are bonded by intermolecular forces. A liquid always takes up the shape of its container.

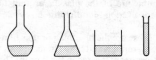

The particles are not fixed in a rigid *lattice* as in a **solid**. During **evaporation** and **boiling** the particles get enough energy to overcome the intermolecular forces and become a **gas**.

Litmus This is a dye extracted from lichen. It is used as an acid–base **indicator**. See illustration overleaf.

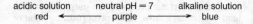

| acidic solution | neutral pH = 7 | alkaline solution |
| red ← | purple | → blue |

Liver This is the largest organ of the vertebrate body and occupies much of the upper part of the abdomen where it is in close association with the **alimentary canal**. The liver is an important organ and has many functions. Some of the main ones are listed below.

(a) Production of bile for use in digestion.

(b) Deamination of proteins which are excess to the needs of the body.

(c) Regulation of blood sugar by the interconversion of **glucose** and **glycogen**.

(d) Storage of iron and **vitamins** A and D.

(e) Detoxification of poisonous substances.

(f) Release and distribution of heat production by the chemical activity of liver cells.

(g) Conversion of stored fat for use by the tissues.

(h) Manufacture of the **plasma protein** fibrinogen which is involved in **blood clotting**.

Lungs These are the breathing **organs** of mammals, amphibians, reptiles and birds. In mammals the lungs are two elastic sacs located in the thorax. They can be expanded or compressed by movements of the thorax in such a way that air is continually taken in (inhaled) and expelled (exhaled).

Air is taken into the body through the windpipe (trachea). The trachea divides into two bronchi which enter the lungs. Each bronchus then further divides into smaller brochioles which eventually terminate as millions of air-sacs called alveoli. The alveoli are richly

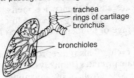

Air passages in the lungs

trachea
rings of cartilage
bronchus
bronchioles

supplied with **blood capillaries** and **gases** are able to diffuse into and out of the bloodstream. See **Breathing in mammals, Gas exchange**.

Machine A device in which an input **force** applied at one point produces an output force elsewhere. Examples of simple machines are **levers, pulleys** and **hydraulic machines**.

Mass This is a measure of the amount of matter in an object. It depends on the number of molecules it contains. The **SI unit** of mass is the kilogram, kg. The mass of an object does not vary with changes in gravity so an object would have the same mass on earth as on the moon.

Mass should not be confused with **weight**, which changes with changes in gravity.

Melting point The temperature at which a solid changes to a liquid (or a liquid changes to a solid). More precisely, it is the temperature at which solid and liquid forms of the same substance (e.g. ice/water) are in equilibrium. At constant pressure the melting point is constant for a pure substance, but it is lowered if impurities are added, hence ice melts when salt is sprinkled on it. The **energy** needed to cause melting without change of temperature is called **latent heat**.

Menstrual cycle This is a reproductive cycle occurring in female primates (monkeys, apes and humans). The cycle is controlled by **hormones**. In human females, the cycle lasts about twenty-eight days. During this time the **uterus** wall thickens in preparation for implantation of a fertilized **ovum**. If fertilization does not occur, the new uterus lining and the unfertilized ovum are expelled from the vagina. This part of the cycle is called *menstruation* and results in a small amount of bleeding from the vagina. See **Fertilization in humans, Ovulation**.

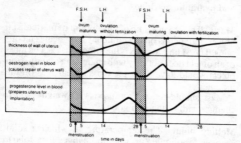

Mercury Mercury is unusual in that it is the only **liquid metal** at room **temperatures**. It has many important uses. Here are some examples:

(a) **Alloys** of mercury with other metals are called amalgams. Zinc amalgam is used for teeth fillings.

(b) Mercury is used as the liquid in thermometers.

(c) In mercury cells, the metal forms the cathode (negative **electrode**) in the **electrolysis** of sodium chloride solution to form sodium hydroxide.

Both mercury vapour and mercury compounds are poisonous and must be handled with extreme care.

Metabolism A collective term describing all of the physical and chemical processes occurring within a living organism. These include both the synthesis (anabolism) and breakdown (catabolism) of **compounds**.

The rate of metabolism of a resting animal, measured by oxygen consumption, is known as the basal metabolic rate (BMR). It is the minimum amount of **energy** needed to maintain life and varies within species, age and gender.

Metal Many **elements** are metals. Pure metals are rarely used in modern life. Most metals are used in the form of **alloys**. Typical properties of metals are;

— crystalline structure.

— strong and hard.

- malleable and ductile.
- produce **cations**.
- react with **acids**.
- react with **nonmetals**.
- produce basic oxides.
- good conductors of **heat** and electricity.
- have high **melting** and **boiling points**.

Methane A gaseous alkane with the formula CH$_4$. In the methane molecule the **hydrogen atoms** are arranged in the shape of a *tetrahedron* around the **carbon** atom.

Methane is the main constituent of **natural gas** and is also released from **petroleum**. It burns readily in a plentiful supply of air to give **carbon dioxide** and water, and is also an industrial source of hydrogen.

Microscope An optical instrument which gives a large image of a small object. In its simplest form a microscope consists of a single converging **lens**. Compound microscopes have two or more lenses. The lens nearest the object is called the *object lens* (objective) while the lens nearest the eye is called the *eye lens* (eyepiece).

Microwaves A region of the **spectrum** of **electromagnetic waves**. The approximate **wavelength** range is 1 mm to 10 m and the approximate **frequency** range is $10^7–10^{12}$ Hz.

Uses of microwaves include microwave cookers and **radar**. In microwave cookers the frequency used (typically 2450 MHz) is strongly absorbed by the food causing it to heat up rapidly. The microwaves penetrate into the food and cook it more quickly than a conventional oven can.

Mineral salts Components of soil formed by the weathering of rocks and from **humus** mineralization. Mineral salts are found in solution in soil water. They are absorbed by plant roots and transported through the plant in solution. Like **vitamins**, mineral salts are needed in tiny amounts but are nevertheless vital for plant and, ultimately, animal nutrition. The absence of a particular mineral salt can lead to mineral deficiency, disease and death. Plants need at least twelve mineral salts for healthy growth.

(a) *Major elements* – needed in relatively large amounts: nitrogen, phosphorus, sulphur, potassium, calcium, magnesium.

(b) *Minor elements* – needed in very small amounts: manganese, copper, zinc, iron, boron, molybdenum. Some mineral salts are needed by plants, some are needed by animals, and some are needed by both. The properties of some important mineral salts are summarized in the table overleaf.

Mineral salt	Function	Some deficiency effects
Calcium	Component of plant cell walls and animal bones.	Rickets in humans.
Iron	Component of haemoglobin.	Anaemia in humans.
Magnesium	Component of **chlorophyll**.	Pale yellow plant **leaves** (chlorosis).
Nitrogen	Component of **protein** and nucleic acids.	Poor reproductive development in plants.
Phosphorus	Components of ATP, nucleic acids, cell membrane, animal bones.	Stunted plant growth.

Mirror An optical device which bends light by **reflection**.

Mixture When two or more substances are present in the same container, but are not chemically bonded together, they are said to be a mixture (see **Compound**). Some properties of mixtures:

(a) The proportions of substances in a mixture can vary.

(b) No **energy** is absorbed or released when a mixture is made.

(c) The **properties** of a mixture are the properties of all the components.

(d) Mixtures can be separated by physical methods.

Mole A mole is the mass of an element or compound which contains the same number of atoms or molecules as there are atoms in 12 g of carbon.

The mass of one mole of an element is the **relative atomic mass** (expressed in grams). The mass of one mole of a **compound** is the **relative molecular mass** (expressed in grams).

Symbol	H	C	O
Relative atomic masses	1	12	16
Mass of 1 mole of atoms	1 g	12 g	16 g
Formula	H_2	C	O_2
Mass of 1 mole of molecules	2 g	12 g	32 g

Thus 1 mole of water molecules is $2 \times 1 + 16 = 18$ g, and 1 mole of carbon dioxide molecules is $12 + 2 \times 16 = 44$ g.

Molecular formula The molecular formula of a substance shows the number and types of **atoms** found in the **molecule**. However, it does not tell us anything about the way in which those atoms are arranged (structure). For example, the molecular formula of water is H_2O, thus a molecule of water consists of two atoms of **hydrogen** and one atom of **oxygen**. It is not unusual for two substances with completely different structures and chemical properties to have the same molecular formula. For example, the chemicals butanol and ethoxyethane both have the molecular formula $C_4H_{10}O$. See **Empirical formula, Structural formula**.

Molecule This is the smallest particle of an **element**

or **compound** which can exist independently. It contains two or more **atoms** bonded together in small whole numbers, e.g.

$$O_2 \quad \text{a molecule of oxygen}$$
$$CH_4 \quad \text{a molecule of methane}$$
$$CaCO_3 \quad \text{a molecule of calcium carbonate}$$

Motor, electric A device which converts input electric **energy** into movement. In most electric motors movement is due to a central rotating coil. The structure of a simple motor is identical to that of a simple **generator**; however, they have opposite effects:

(a) In a generator the coil is turned and an electric current is produced in it.

(b) In an electric motor an electric current passes through the coil and causes it to turn.

Mucus A slimy fluid secreted by goblet cells found in the epithelia of vertebrates. Mucus has several important functions in humans.
(a) It traps dust and bacteria in the air passages.
(b) It lubricates the surfaces of the internal organs.
(c) It lubricates the inside walls of the **alimentary canal** making it possible for food to pass through easily.
(d) The layer on the inside walls of the alimentary canal prevents digestive **enzymes** from reaching and digesting the walls themselves.

Multicellular (of an organism) This word describes

an organism which consists of many cells. Most animals and plants are multicellular. Compare **Unicellular**.

Muscle Animal **tissue** which consists of **cells** which are able to contract in response to nerve impulses, thus producing movement. This may be movement of internal **organs** within the animal or movement of the animal in its environment.

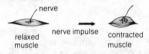

relaxed muscle   nerve impulse   contracted muscle

Mutation This describes a natural spontaneous change in the structure of **DNA** in **chromosomes**. Mutations are rare, but when they occur the resulting altered characteristic is passed on to subsequent generations. Mutations most often confer disadvantages on the organism inheriting them; however, mutations may also result in beneficial **variations** within a **population**. Such beneficial variations are the basis of **evolution**.

Mutations can be induced in organisms by exposure to excessive **radiation**. The artificial alteration of DNA in chromosomes is called *genetic engineering*. Scientists are currently looking at ways of using this technique to improve the yield and immunity to disease of many food crops.

Natural frequency Any system which can vibrate (see **Vibration**) will vibrate freely at one particular frequency, its natural frequency. The value of this depends on the physical nature of the system. For example, a simple **pendulum** will swing at a **frequency** which depends on its length and on the gravitational field strength at that place. The frequency of the note produced by blowing across the top of a bottle depends on the size and shape of the air space inside it. These examples of **resonance** are matched by others in electronics, radiation and nuclear physics.

Natural gas A **mixture** containing mainly **hydrocarbon gases** which is found in deposits beneath the earth's surface. Natural gas and **petroleum** are often found together. **Methane** is usually the major constituent and often makes up over 90% of the mixture. The **noble gas** *helium* is sometimes present in small amounts. Natural gas is used as a fuel both in industry and the home.

Natural selection This is a theory first proposed by the famous scientist Charles Darwin to suggest how **evolution** could have taken place. Darwin suggested that individual organisms within a **species** differ in the extent to which they are adapted to conditions of their environment. Thus, in competition for food etc., the better-adapted organisms will survive and pass on their favourable **variations** to future generations. Conversely, the less well-adapted will die out.

Nervous system A network of specialized cells in *multicellular* animals. This network links receptors and effectors, thus coordinating the animal's activity. In mammals, the nervous system consists of the brain and the **spinal cord** (which together are called the central nervous system) and nerve cells connecting to all parts of the body.

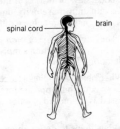

spinal cord — brain

Neutralization In this process either the **pH** of an **acid** solution is increased to 7 or the pH of an alkaline solution is decreased to 7. pH 7 is neutral. An acid can be neutralized by adding a base or a **compound** such as a *carbonate*.

$$HCl(aq) + NaOH(aq) \rightarrow NaCl(aq) + H_2O(l)$$
$$2HCl(aq) + CuCO_3(s) \rightarrow CuCl_2(aq) + CO_2(g) + H_2O(l)$$

In acid-base reactions, **indicators** are used to show when the neutralization has occurred.

Neutron One of the three main *subatomic particles* found in most **atoms**. The neutron has the same **mass**

as a **proton** but carries no electrical **charge**.

Newton's laws of motion   These are three important statements which relate **forces** to each other and their effects on objects. They are valid for all forces and all cases, and form the basis of much physics.

First law: A body stays at rest, or if moving it continues to move with uniform **velocity**, unless an external force acts on it making it behave differently.

Second law: The rate of change of *momentum* equals the net force. This is often written simply as; Force = mass × acceleration ($F = m \times a$).

Third law: When object A applies a force on object B, then B applies an equal but opposite force on A. This is often expressed as action and reaction are equal and opposite.

The second law helps to define the newton, N the unit of force. It is the force which will accelerate a mass of 1 kg by 1 m/s^2.

Nitrogen cycle   The circulation of the element *nitrogen* and its **compounds**. Much of this cycle is the result of metabolic processes (see **Metabolism**) of living organisms. However, there are large amounts of synthetic nitrogen-containing **fertilizers** made from atmospheric nitrogen (via **ammonia** and nitric acid), used each year. These also play a significant part. Nitrates may also result from the action of lightning on atmospheric gases. The following diagram summarizes the nitrogen cycle.

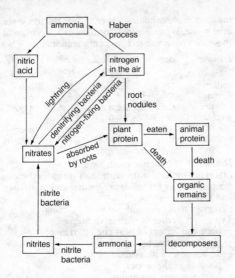

Noble Gases (or Inert Gases) These are **elements** in **Group** 0 of the **periodic table**. They are all nonatomic gases found in small amounts in the atmosphere, i.e. helium, neon, argon, krypton, xenon.

Their electronic configurations are very stable because they have a complete outer shell of **electrons**. The elements do not easily lose or gain electrons. A few chemical **compounds** exist which contain atoms of the more massive noble gases (e.g. XeF_4) but these are man-made. When elements form **ions**, the electrons of these ions have an inert gas configuration, e.g.

Element	Ion	Electronic configuration atom	ion	Noble gas
sodium	Na^+	2.8.1	2.8	neon
fluorine	F^-	2.7	2.8	neon
calcium	Ca^{2+}	2.8.8.2	2.8.8	argon
sulphur	S^{2-}	2.8.6	2.8.8	argon
aluminium	Al^{3+}	2.8.3	2.8	neon
bromine	Br^-	2.8.18.7	2.8.18.8	krypton

Nonmetal **Elements** which have:
(a) either molecular structures and are gases at room temperature, or are solids or liquids with low **melting** and **boiling points** or
(b) *giant structures* with **covalent bonding**.
Typical properties of nonmetals are;
— poor conductors of heat and electricity.
— give rise to **anions**.
— do not react with **acids**.
— produce acidic oxides.
— form **covalent compounds**.

Nuclear power The supply of useful **energy** from nuclear **fission** reactions in a *nuclear reactor*. The energy is produced by a controlled **chain reaction**. The **kinetic energy** of the particles produced by each fission produces a rise in temperature. A coolant removes the heat energy to make steam. The steam drives turbines which, in turn, drive **generators**.

Nucleus **1.** The part within most cells which contains **chromosomes**. As the chromosomes contain the hereditary information, the nucleus controls the cell's activity through the action of the genetic material, **DNA**. The nucleus is isolated from the *cytoplasm* by a *nuclear membrane*.

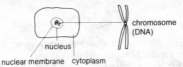

2. This is the part at the centre of an atom where the mass is concentrated. It contains **protons** and **neutrons**. The nucleus has a positive **charge**, and in the neutral atom this is balanced by negatively charged **electrons** moving around the nucleus. Atoms of the **isotope** hydrogen-1 are the only atoms whose nuclei contain no neutrons.

The nucleus of a lithium atom

Nylon The nylons are a group of *polyamide polymers*. The commonest of them is called Nylon 6.6 (see diagram overleaf).

This *man-made fibre* has great strength. It is used in fabrics, yarns (stockings, knitwear), carpets, ropes and nets. Nylon is useful because it will not rot, is hard-

$$
\begin{array}{c}
\text{—N—C—CH}_2\text{—CH}_2\text{—CH}_2\text{—CH}_2\text{—C—N—CH}_2\text{—CH}_2\text{—CH}_2\text{—CH}_2\text{—CH}_2\text{—CH}_2\text{—N—C—}
\end{array}
$$

6 carbon atoms   6 carbon atoms

Nylon 6.6

wearing and does not absorb water; however, it does stretch (useful in ropes and stockings).

Nylon is often mixed with other fibres such as wool to get a balance of properties. Generally speaking a nylon jumper is not as warm as a woollen one, however it will not wear out as quickly. A mixture of fibres produces a jumper which is both warm and hard-wearing.

Oestrogen A **hormone** which is secreted by the **ovaries** of vertebrates. It stimulates the development of **secondary sexual characteristics** in female mammals, and plays an important role in the **menstrual cycle**.

Oil A **liquid fat** such as melted butter, olive oil, sunflower oil, etc. The word is sometimes used in place of **petroleum**.

Oleum A solution of sulphur (VI) oxide (sulphur trioxide) in concentrated **sulphuric acid**. Oleum is extremely corrosive and is a vigorous oxidizing agent, see **Oxidation**.

Optic nerve A cranial nerve found in vertebrates. It conducts nerve impulses from the retina of the eye to the brain.

Orbit The closed path of an object moving in a central force field. For example, the moon travels in an orbit around the earth moving in the earth's gravitational field. The central force field is continually pulling the object towards it, thus preventing it from moving in a straight line. The object has both **potential** and **kinetic energy**.

Ore A naturally occurring substance from which an **element** (often a **metal**) can be extracted. For example:

Ore	Element
galena	lead
haematite	iron
calomine	zinc
bauxite	aluminium
pitchblende	uranium

Organ A collection of different **tissues** in a plant or animal which forms a structural and functional unit within an organism, such as the **leaf** of a plant or the **liver** of an animal. Different organs may be associated together to constitute a system; for example, the digestive system.

cells → tissues → organs → systems

Organic chemistry This branch of chemistry is concerned with the study of **compounds** of **carbon** but does not include carbonates, hydrogencarbonates, **carbon dioxide**, etc. Compare **Inorganic chemistry**.

Oscillator A **circuit** whose output is an **alternating current**. It is normal to be able to *tune* the **frequency** of the output by changing the value of a *capacitor* in the circuit. The active component in most oscillators is a *valve* or **transistor** which acts as an **amplifier**. Part of the output energy from the amplifier is fed back into the input to control the action. This is called *feedback*.

Osmoregulation The control of the osmotic **pressure** and therefore the water content of an organism (see **Osmosis**). Here are some examples of osmoregulation.

(a) *Seawater organisms.*

low salt and high water concentration

continual swallowing of water

high salt and low water concentration

water leaves by **osmosis**

(b) *Freshwater organisms.*

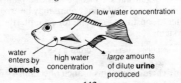

low water concentration

water enters by **osmosis**

high water concentration

large amounts of dilute **urine** produced

(c) *Terrestrial organisms*. Water is gained from food and drink and as a by-product of **respiration**. Water is lost (i) during sweating, (ii) in exhaled air (iii) as urine. The balance of water and mineral salts in the bodies of terrestrial animals is mainly controlled by the kidneys.

Osmosis The **diffusion** of a solvent (usually water) through a selectively permeable membrane from a region of high solvent **concentration** to a region of lower solvent concentration. For example:

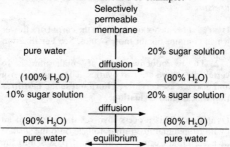

Selectively
permeable
membrane

pure water	20% sugar solution
	diffusion →
(100% H$_2$O)	(80% H$_2$O)
10% sugar solution	20% sugar solution
	diffusion →
(90% H$_2$O)	(80% H$_2$O)
pure water	← equilibrium →	pure water

Examples of selectively permeable membranes are:

(a) The **cell** membrane.

(b) Visking (dialysis) tubing.

Selectively permeable membranes are thought to have tiny pores which allow the rapid passage of small water **molecules** but restrict the passage of larger solute molecules.

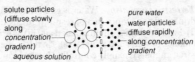

solute particles
(diffuse slowly along concentration gradient)

aqueous solution

pure water

water particles diffuse rapidly along *concentration gradient*

selectively permeable membrane

Since the cell membrane is selectively permeable, osmosis is important in the passage of water into and out of cells and organisms. The pressure exerted by the movement of water due to osmosis is called *osmotic pressure*.

Ovary **1.** The hollow region in the **carpel** of a flower which contains one or more ovules. See **Fertilization in plants**.
  2. The reproductive **organ** of female animals. In vertebrates there are two ovaries. They produce **ova** and also release certain sex **hormones**. See **Fertilization in humans, Ovulation**.

Ovulation This process involves the release of an **ovum** from a mature graffian follicle onto the surface of a vertebrate **ovary**. From here it passes into the oviduct and then into the **uterus**.
  In the human female, ovulation is controlled by **hormones** secreted by the pituitary gland. The sequence of events in the female's reproductive behaviour is called the **menstrual cycle**.

Ovum An unfertilized female **gamete** produced at the **ovary** of many animals. An ovum contains a *haploid* **nucleus**. See **Fertilization, Ovulation**.

Oxidation A substance is oxidized if it:

(a) gains **oxygen**, e.g.

$$2Mg(s) + O_2(g) \rightarrow 2MgO(s)$$
magnesium   oxygen   magnesium oxide

(b) loses **hydrogen**, e.g.

$$CH_4(g) + Cl_2(g) \rightarrow CH_3Cl(l) + HCl(g)$$
methane   chlorine   chloromethane   hydrogen
                                        chloride

(c) loses **electrons**, e.g.

$$Cu(s) \rightarrow Cu^{2+}(aq) + 2e^-$$
copper   copper ions   electrons

A substance which brings about the oxidation of another substance is called an oxidizing agent. There are several common oxidizing agents.

oxygen	O_2
chlorine	Cl_2
ozone	O_3
hydrogen peroxide	H_2O_2
potassium manganate (VII)	$KMnO_4$

Compare **Reduction**.

Oxygen (O_2) A gaseous nonmetallic element in **group** 6 of the **periodic table**. The gas is both colourless and odourless and makes up 21% of the **atmosphere**. It is vital for the **respiration** of both plants and animals. Although atmospheric oxygen is used up during respiration the supply is continually replenished by plants as a result of **photosynthesis**. Oxygen is only slightly soluble in water; however, there is normally enough dissolved to support fish and other aquatic life. Pollution often causes a drop in the **concentration** of oxygen in the water to such an extent that fish will suffocate and die. Oxygen is obtained industrially by the *fractional distillation* of liquefied air. In the laboratory it is obtained by the decomposition of hydrogen peroxide, usually with the help of a **catalyst** such as manganese (IV) oxide.

$$2H_2O_2(aq) \rightarrow 2H_2O(l) + O_2(g)$$

The chemical test for oxygen is to put a glowing splint into the gas. If the gas is oxygen the splint will relight. Oxygen will also make anything which is already burning in air burn much more fiercely.

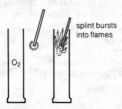

splint bursts into flames

O_2

Oxygen is a vigorous *oxidizing agent*. It is used extensively in steel-making and welding, as a rocket propellant together with kerosine or hydrogen, and for life-support systems in medicine.

Ozone layer   The gas ozone is an allotrope of **oxygen** where the molecule is a triatomic (O_3). The ozone layer is a region of the **atmosphere** about 30 km above the earth's surface, which is rich in ozone. This layer prevents most of the **ultraviolet** radiation from reaching the earth. This is important, as exposure to excess ultraviolet radiation is known to cause skin cancer.

Scientists have recently become aware of holes appearing in the ozone layer allowing unusually large amounts of ultraviolet radiation to reach the earth. Chemicals called *chlorofluorocarbons (CFCs)*, used as aerosol propellants and in a wide range of industrial processes, are thought to be the cause. Some CFCs eventually diffuse up into the ozone layer. They are very unreactive chemicals under normal room conditions; however, in the presence of sunlight they react with the ozone thus destroying it. Steps are now being taken to reduce the amounts of CFCs used in the world.

Pancreas   This is a gland situated near the duodenum (see **Alimentary canal**) of vertebrates. It releases an alkaline fluid containing digestive **enzymes** such as lipase, amylase and trypsin into the duodenum (see **Digestion**). The pancreas also contains **tissue** known as the *Islets of Langerhans* which secretes the hormone insulin, which is important in the metabolism of sucrose.

Parasite   An organism that feeds in or on another living organism called the *host*. In a parasitic relationship the host does not benefit and may actually be harmed.

Examples of parasites in man include fleas, lice and tapeworms.

Particle In physics this word is used to describe an object whose **volume** is very small. The **kinetic model** describes the structure of matter in terms of particles. In simple pictures of matter, such particles are called **atoms** or **molecules**. Studies have revealed that particles can sometimes appear as **waves** because they can show wave properties like **diffraction** and **interference**.

Pascal (Pa) The unit of pressure. 1 pascal = 1 newton/metre².

Pendulum A regularly swinging object with a regular transfer to and fro between **potential** and **kinetic energy**.

	A	B	C
potential energy	yes	no	yes
kinetic energy	no	yes	no

The period of a pendulum is the time it takes for one full cycle; in this case to swing from A to C and back to

148

A. The value of the period depends only on the length, l, of the pendulum and the acceleration due to gravity, g. The value of g can be estimated by carrying out a simple experiment with a pendulum and putting the results into the following equation.

$$g = \frac{4 \times \pi^2 \times l}{T^2}$$

Periodic table This is a way of representing all of the elements so as to show similarities and differences in their chemical properties. See table overleaf.

The elements are arranged in increasing order of **atomic number (Z)** as you read from left to right across the table. The horizontal rows are called *periods* and the vertical columns are called **groups**. There is a progression from **metals** to **nonmetals** across the period. This is period 3:

element	Na	Mg	Al	Si	P	S	Cl	Ar
atomic number	11	12	13	14	15	16	17	18
		metals					nonmetals	

This is part of group 1;

Li
Na
K

The block of elements between Groups 2 and 3 are called *transition metals*. The chemistry of these metals is similar in many ways. Elements 58 − 71 are known as the *Lanthanide* or *Rare Earth* elements. Elements 90 − 103 are known as the *Actinide* elements. The ele-

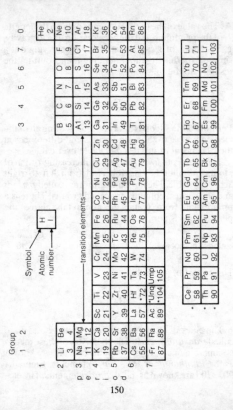

150

ments which have atomic numbers greater than 92 do not occur naturally but have been made artificially by bombarding other elements with particles.

Pest A general term used to describe any organism which is considered to have a detrimental effect on humans. This effect may be to the body, to the food supply or to the environment. Some examples are shown in the following table.

Example of pest	Effect on humans
weeds; locusts	Reduces the growth of plants and crops.
foot and mouth virus	Causes disease in domestic animals.
woodworm, wet rot fungus	Damages buildings.
mosquito; lice	Transmits human disease.

Methods used to combat pests include:
(a) Spraying with chemical **(pesticides)**.
(b) Using natural **predators** against the pest.
(c) Introducing **parasites** and **pathogens** to the pest **population**.
(d) Introducing sterile (unable to breed) individuals to the pest population, thus reducing the reproductive capacity.

Pesticide A chemical **compound**, often delivered as a spray or fine powder, which kills or prevents the growth of pests which damage crops. Pesticides are often divided into three groups:

(a) *Herbicides.* Chemicals which act on plants (weed-killers), e.g. paraquat.

(b) *Fungicides.* Chemicals which act on fungal growths, e.g. cheshunt compound.

(c) *Insecticides.* Chemicals which act on insects, e.g. DDT (now banned in Britain).

Pesticides are used to increase the yield of crops, however there are certain disadvantages associated with their use:

(a) They may kill organisms other than the target pest.

(b) The concentration of pesticide increases as it passes through a **food chain**.

(c) Some pesticides decompose very slowly and may accumulate into harmful doses within consumer organisms.

(d) By killing off susceptible organisms, they allow resistant individuals to grow and multiply with reduced competition.

Petroleum   This term describes the **mixture of hydrocarbons**, e.g. **natural gas** and *crude oil*, found in the earth's crust. It is thought to have been produced over millions of years by the action of heat and pressure on the remains of marine animals and plant organisms.

There are large reserves of petroleum in the Middle East, United States, Soviet Union, Central America and North Sea. Petroleum is the raw material of the petrochemical industry and is the source of petrol, diesel fuel, heating oil, fuel oil and gas supplies. These products are obtained by the *fractional distillation* of petroleum. This method of separation is possible because the different components, or fractions, of the mixture boil at different temperatures. The diagram overleaf shows how petroleum might be fractionally distilled.

In recent years petroleum has replaced **coal** as the major source of raw materials for the chemical industry. However, scientists are aware that the supply of petroleum cannot last forever and the search is now on to find alternative sources.

pH The pH scale is a measure of the *acidity* or *alkalinity* of a solution. The lower the value, the more acidic the solution, i.e. the higher the **concentration** of hydrogen ions (H^+) it has. A neutral solution has equal concentrations of hydrogen and **hydroxide (OH^-) ions** and has a pH of 7 at 25 °C.

increasing acidity   neutral   increasing alkalinity

0 1 2 3 4 5 6 7 8 9 10 11 12 13 14

Photosynthesis The process by which green plants make **carbohydrate** from **carbon dioxide** and water. The energy for the reaction comes from sunlight which is absorbed by the green pigment **chlorophyll** found in the **chloroplasts.** Oxygen is produced as a byproduct in this process.

153

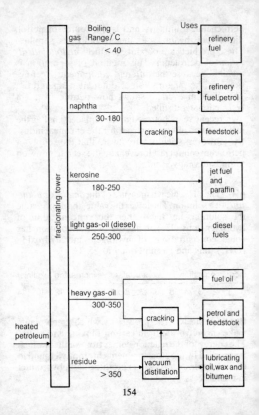

	Boiling Range/°C		Uses
gas	< 40		refinery fuel
naphtha	30-180		refinery fuel, petrol
		cracking	feedstock
kerosine	180-250		jet fuel and paraffin
light gas-oil (diesel)	250-300		diesel fuels
heavy gas-oil	300-350		fuel oil
		cracking	petrol and feedstock
residue	> 350	vacuum distillation	lubricating oil, wax and bitumen

heated petroleum

fractionating tower

$$\text{carbon dioxide} + \text{water} \xrightarrow[\text{chlorophyll}]{\text{light energy}} \text{carbohydrate} + \text{oxygen}$$

$$6CO_2 \qquad 6H_2O \qquad \qquad C_6H_{12}O_6 \ + \ 6O_2$$

Photosynthesis is the source of all food and the basis of **food chains**, while the release of oxygen replenishes the oxygen content of the atmosphere. The process is thus essential to the functioning of the **biosphere**.

Physical change A physical change to a substance involves changes in its physical properties, e.g. temperature particle size, state, etc., but no alteration to its chemical properties. Compare **Chemical change**.

Placenta This organ develops in the **uterus** of a female mammal during **pregnancy**. It allows a close association between the blood circulation of the mother and that of the **foetus**. The placenta allows the passage of food and oxygen from the mother to the foetus and the removal of the waste products **carbon dioxide** and **urea** from the foetus.

Plankton The collective word used for microscopic animals and plants which float in the surface waters of lakes and seas. Plankton play a vital role as the starting point of aquatic **food chains**.

Plasma The clear fluid of vertebrate blood in which the blood cells are suspended. It is an aqueous solution which contains many dissolved **compounds** being transported to different parts of the body, e.g.

| carbon dioxide | }| waste products; |
| urea | | |

glucose	}	digested foods;
amino acids		
hormones		
plasma proteins		
sodium chloride		

Pollen The reproductive spores of flowering plants. Each pollen grain contains a male **gamete**. Pollen grains may be transferred either by insects or by the wind, and they are adapted to their mode of transfer.

Insect pollination   Wind pollination
spiky and sticky   smooth and light

air bladders

See **Fertilization in plants, Pollination**.

Pollination The transfer of **pollen** grains from the **stamens** (male part of a flower) to **carpels** (female part of a flower) in flowering plants.
(a) *Self-pollination* involves the transfer of pollen within the same flower or between flowers on the same plant.
(b) *Cross-pollination* involves the transfer of pollen between two separate plants.

Normally the male and female parts of a plant do not mature at the same time so crosspollination is more likely. This results in the mixing of **chromosomes** and may lead to **variation**. Pollen is transferred either by the wind or on the bodies of insects. Flowers are adapted to favour one particular method of pollen transfer.

(a) *Insect pollination* − the flowers are generally highly coloured and/or scented and contain nectar. Insects are attracted by the colour and scent and visit the flower to collect the nectar. Their bodies become dusted with pollen. Some of this pollen may stick to the stigmas of subsequent flowers which they visit.

(b) *Wind pollination* − the flowers are often drab and have no scent. They produce many more pollen grains than insect-pollinated plants, as much of the pollen is lost in transfer. See **Fertilization in plants, Flowers, Pollen**.

Pollution A general word for the addition of any substance to an **environment** which upsets the natural balance. Pollution has resulted mainly from *industrialization*. Two important factors have been a large increase in the **combustion** of **fossil fuels** and a migration of people from the land to towns and cities. Pollution of the air and water are often singled out as particularly important (see **Biosphere**).

(a) Air pollution is caused mainly by the combustion of fossil fuels.

coal → smoke + carbon dioxide + sulphur dioxide
petrol → smoke + carbon + nitrogen + **lead**
                monoxide oxides

Air pollutants such as smoke and sulphur dioxide cause irritation to the human respiratory system and may accelerate diseases such as bronchitis and lung cancer. Sulphur dioxide and nitrogen oxides are responsible for **acid rain**. Lead is thought to retard the mental development of children.

(b) Water pollution is caused by the intentional or accidental addition of materials into both freshwater and seawater. Most pollutants originate from industrial and agricultural practices and from the home. For example, mine and quarry washings, **acids, pesticides, fertilizers**, oil, radioactive discharges, detergents, hot water (from power stations). Some pollutants, such as pesticides, may poison and kill aquatic organisms. Other pollutants, such as sewage, cause an increase in the populations of microorganisms in the water. The result is a reduction in the dissolved oxygen level making it impossible for other organisms to survive. This process is called *eutrophication*.

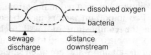

Polymers Polymers are large **molecules** in which a group of atoms is repeated; e.g.

$$x - x - x - x - x - x - x - x$$
$$\text{or}$$
$$x - y - x - y - x - y - x - y$$

Polymers may be naturally occurring substances such as **starch** and **cellulose**, or man-made substances such as **nylon** and **polythene**. Polymers are made by reacting *monomer* molecules together, usually in the presence of a catalyst. Here are some examples of common polymers.

monomer	polymer

$CH_2 = CHCl$
chloroethene
(vinyl chloride)

$-CH_2-\overset{\displaystyle H}{\underset{\displaystyle Cl}{C}}-\left[CH_2-\overset{\displaystyle H}{\underset{\displaystyle Cl}{C}}\right]_n-CH_2-\overset{\displaystyle H}{\underset{\displaystyle Cl}{C}}-$
polychloroethene
(polyvinyl chloride PVC)

$\overset{\displaystyle H}{\underset{\displaystyle H}{C}}=\overset{\displaystyle H}{\underset{\displaystyle H}{C}}$
ethene

$-CH_2-\overset{\displaystyle H}{\underset{\displaystyle H}{C}}-\left[CH_2-\overset{\displaystyle H}{\underset{\displaystyle H}{C}}\right]_n-CH_2-\overset{\displaystyle H}{\underset{\displaystyle H}{C}}-$
polythene

phenylethene C_8H_8
(styrene)

polyphenylethene
(polystyrene)

Potential energy This term is usually used for the **energy** which an object has stored up because of its position. For example, water at the top of a waterfall has potential energy. As the water drops, the potential energy changes into **kinetic energy**. Potential energy can also be stored as elastic, chemical or electrical energy.

Power (P) The rate of **energy** transfer or the rate or working. The unit, the **watt**, is the transfer of one **joule** per second. The horsepower was once used as a unit of power; one horsepower is equivalent to just under 750 watts. Here are some useful equations involving power.

$$\text{mechanical power} = \text{force} \times \text{velocity} = \frac{\text{work done}}{\text{time taken}} = Fv$$

$$\text{electrical power} = \text{voltage} \times \text{current} = VI$$

Efficiency is often expressed in terms of power.

$$\text{efficiency} = \frac{\text{useful output power}}{\text{total input power}}$$

In the case of a **transformer**;

$$\text{efficiency} = \frac{P_2}{P_1} = \frac{V_2 \times I_2}{V_1 \times I_1}$$

Precipitate (ppt) The insoluble substance formed when two solutions are mixed in a double decomposition reaction. In each of the following examples the precipitate is underlined and appears as a solid when the two clear solutions are mixed.

$$Pb(NO_3)_2(aq) + 2NaCl(aq) \rightarrow PbCl_2(s) + 2NaNO_3(aq)$$

| lead (II) nitrate | sodium chloride | lead (II) <u>chloride</u> | sodium nitrate |

$$BaCl_2(aq) + Na_2SO_4(aq) \rightarrow BaSO_4(s) + 2NaCl(aq)$$

barium chloride	sodium sulphate	barium <u>sulphate</u>	sodium chloride

Predator An animal which feeds on other animals which are called the *prey*. A predator is a food consumer but not a **parasite**. The relationship between predator and prey is reflected in their numbers. The following diagram shows how the numbers of predators and prey in an area may follow a cycle.

We can interpret the cycle as follows;
(a) When prey is plentiful a larger number of predators will survive as there is less competition for food.
(b) More predators will eat more prey so the number of prey will decline.
(c) Fewer prey enable fewer predators to survive as there is now more competition for food.
(d) Fewer predators will eat fewer prey, so the number of prey will increase.

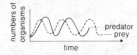

Pregnancy (or Gestation period) In mammals this signifies the time between conception and birth. Human pregnancy lasts for approximately forty weeks and during this time the **embryo**, which has become implanted in the **uterus**, develops. In the early stages of pregnancy, finger-like structures (called *villi*) grow

161

from the embryo into the uterus wall and develop into the **placenta**. This allows the transfer of oxygen and food from the mother to the embryo and the removal of **carbon dioxide** and **urea**.

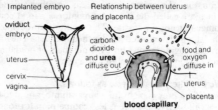

During the pregnancy the cells of the embryo continually divide and differentiate. The growing embryo (now called a **foetus**) becomes suspended in a water-filled sac called the *amnion*. The placenta extends into the umbilical cord which connects with the abdomen of the foetus.

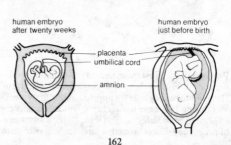

Pressure (*p*) The **force** on a unit area of a surface from the substance in contact with it. The unit of pressure is the pascal (Pa) (1 Pa = 1 N/m²). To calculate pressure we use the equation; pressure (Pa) = force (N)/area (m²). Often the force applied to an object is weight *W*.

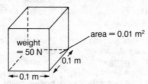

pressure exerted on the ground $= \dfrac{50}{0.01} = 5\,000$ Pa

The pressure at a depth in a liquid or in the air equals the weight above the unit area. This gives the relationship; pressure = depth × mean **density** × *g*. The pressure exerted by the air is called *atmospheric pressure*. Pressure may be measured by a variety of instruments, e.g. manometer, Bourdon gauge, barometer.

Primary sexual characteristics These are the features which distinguish between males and females from the time of their birth. Primary sexual characteristics do not include those which develop at puberty and are characteristic of adulthood. Compare **Secondary sexual characteristics**.

Progesterone This is a hormone secreted by the ovaries of a mammal. It prepares the **uterus** for **implanta-**

163

tion and prevents any further **ovulation** during the **pregnancy**.

Properties The properties of a substance are the characteristic ways in which it behaves (reacts) that make it what it is and make it different from other substances. Properties are often classified as *physical* or *chemical*. Chemical properties relate to the chemical reactions of the substance.

Chemical properties include whether the substance:	Physical properties include:
is a **metal** or **nonmetal**;	colour
gives acidic or basic oxides;	density
has more than one **valency**;	physical state
reacts with **acids**;	boiling point
is an oxidizing or reducing agent	melting point
	solubility

Proteins Organic **compounds** containing the elements carbon, hydrogen, oxygen, nitrogen and sometimes sulphur. A simple molecule of protein consists of a long chain of subunits called **amino acids**. These chains may be joined to other chains and folded in several different ways resulting in very large and complex molecules.

Proteins are the building blocks of cells and tissues, being important constituents of muscle, skin, bone, etc. Proteins also play a vital role as **enzymes**, and some **hormones** also have a protein structure.

Proton A positively charged subatomic particle found in the **nucleus** of the atom. The number of protons in an atom is given by the **atomic number**.

Isotopes of any element always contain the same number of protons.

Pulse rate   The regular beating of blood in **arteries** due to the **heartbeat**. Pulse rate can be detected in the human body where an artery runs close to the skin. The wrist is often used for taking someone's pulse. In an adult human, pulse rate varies between about 70 beats per minute at rest to over 100 beats per minute during exercise.

Radar   A technique for finding the direction and distance of an object by analysing its **reflection** of **microwaves**.

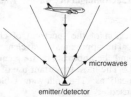

microwaves

emitter/detector

Pulses of waves are emitted in different directions. Waves which hit objects are reflected back to a detector. The direction from which the waves are reflected gives the direction of the object. From knowing the speed of the microwaves, and the time taken to reach the object and be reflected back to the detector, it is possible to calculate the distance to the object.

If the object is moving there is a **frequency** change during reflection; this is called the *Doppler effect*. The size of the frequency change gives the object's speed.

This method is used by traffic police to measure the speed of vehicles.

Radiation **Energy** transfer either by **waves** or **particles**. Any radiating object therefore loses energy. Heat radiation is energy transfer by **infrared** waves. It leaves all objects at a rate that depends on **temperature**, However, black surfaces radiate (and absorb) these waves better than white ones unless the temperature is very high. Cosmic radiation consists of a stream of very fast particles. The source of cosmic radiation is not known for certain, but is thought to be exploding stars. See **Electromagnetic waves, Radioactivity, Sound**.

Radio A region of the **spectrum** of **electromagnetic waves**. The approximate wavelength range is $10 - 10^6$ m and the approximate **frequency** range is $10^2 - 10^7$ Hz. The most important use of radio is in telecommunications. The following table lists the main bands used.

Band	Wavelength range/m
LF low frequency	$10^4 - 10^3$
MF medium frequency	$10^3 - 10^2$
HF high frequency	$10^2 - 10$
VHF very high frequency	$10 - 1$
UHF ultra-high frequency	$1 - 0.1$

Part of the VHF and UHF bands are in the **microwave** region of the electromagnetic spectrum; however, they are often included with radio as they are also used for telecommunication.

Radio communication is helped by the *ionosphere*. This is a region of the upper atmosphere, from about 40 − 400 km above the ground. Many of the air particles in the ionosphere are ionized (hence the name), mainly by **ultraviolet radiation** from the sun. The ionized layer reflects the radio waves.

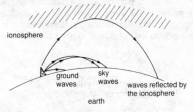

Radio waves which have wavelengths of less than 10 m are used to carry television programmes. Unfortunately these are not reflected by the ionosphere, and communication of live television programmes across the world depends on reflection from suitably placed satellites.

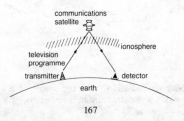

Radioactivity The decay (breakdown) of unstable **nuclei** into more stable forms. During the decay **energy** is transferred to the products; they move away; they have **kinetic energy**. There are several forms of radioactivity.

(a) Alpha decay involves the release of **alpha particles**. The alpha particle, α, is a helium nucleus and has a **mass** of four units. When a parent nucleus, X, changes to a daughter nucleus, Y, the mass of Y is four units less than that of X and its **proton** number is two less.

$$^A_Z X \rightarrow ^{A-4}_{Z-2} Y + ^4_2 He\ (\alpha) + energy$$

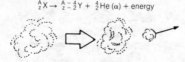

(b) There are several forms of beta decay. In the most common, a **neutron** in the parent nucleus changes to a proton and an **electron beta particle**, ß, escapes.

$$^A_Z X \rightarrow\ _{Z+1}^{A} Y\ +\ _{-1}^{0}e\ (\beta) + energy$$

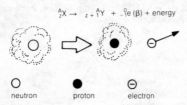

○	●	⊖
neutron	proton	electron

(c) **Gamma radiation**, γ, is the result of a simple change of energy structure in the nucleus.

$$^A_ZX \rightarrow {^A_Z}X + {^0_0}\gamma \text{ (energy)}$$

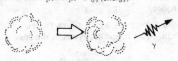

(d) In the case of **fission** the parent nucleus splits into two smaller daughters with the release of a few neutrons. A **chain reaction** may occur.

(e) A few radioactive nuclei become more stable by the release of protons.

In all cases the daughter nuclei may themselves be radioactive, and may produce a different form of radioactivity from the parent nucleus. For this reason it is difficult to provide a pure source of any one form of radioactive radiation. See **Half-life**.

Rate of reaction The rate (or speed) at which a chemical reaction proceeds depends on several factors:

(a) **Temperature**. The higher the temperature the faster the reaction, because the particles are moving with greater energies.

(b) *Particle size.* The smaller the particles involved the greater the total surface area available for reaction, hence the faster the reaction.

(c) **Concentration.** The more concentrated a solution (or the higher the **pressure** of a **gas**) the more particles there are in a given volume. The more particles there are available for reaction, the faster the reaction will be.

(d) **Catalysts**. They change the rate by providing an

alternative reaction pathway along which the reaction can proceed.

Rectifier A device which allows **current** in only one direction. Most forms are types of **diodes**.

conducts

blocks

Rectifiers are used to convert **alternating current** into **direct current**. The following diagrams show the effect which rectification has on alternating current.

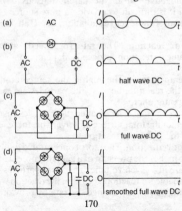

(a) AC

(b) AC DC
half wave DC

(c) AC DC
full wave DC

(d) AC DC
smoothed full wave DC

The smoothing capacitor makes the signal more steady, however the output from (d) may still not be steady enough for all uses. The **Circuit** below shows another simple rectifier. It is the type used in power packs for toy trains and in microcomputers.

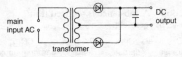

Red blood cell (or **red blood corpuscle** or **erythrocyte**)   The most numerous **cell** in vertebrate **blood**. Red blood cells are responsible for transporting oxygen from the lungs to the **tissues** of the body. In humans, they are biconcave discs and are made in the **bone** marrow. When formed, a red blood cell has a **nucleus**; however, this is lost by the time the cell enters the blood.

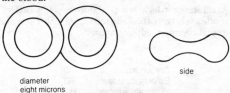

diameter
eight microns

side

Red blood cells contain haemoglobin. When blood passes through the lungs, haemoglobin combines with oxygen to form the unstable **compound** oxyhaemoglobin. At the tissues this compound breaks down releasing oxygen to the cells.

$$\text{haemoglobin} + \text{oxygen} \underset{\text{tissues}}{\overset{\text{lungs}}{\rightleftharpoons}} \text{oxyhaemoglobin}$$

A shortage of red blood cells is called anaemia. Compare **White blood cell**.

Reduction A substance is reduced if it:

(a) loses **oxygen**, e.g.

$$\underset{\text{lead oxide}}{PbO(s)} + \underset{\text{carbon}}{C(s)} \rightarrow \underset{\text{lead}}{Pb(s)} + \underset{\text{carbon monoxide}}{CO(g)}$$

(b) gains **hydrogen**, e.g.

$$\underset{\text{chlorine}}{Cl_2(g)} + \underset{\text{hydrogen}}{H_2(g)} \rightarrow \underset{\text{hydrogen chloride}}{2HCl(g)}$$

(c) gains **electrons**, e.g.

$$\underset{\text{sodium ion}}{Na^+(l)} + \underset{\text{electron}}{e^-} \rightarrow \underset{\text{sodium}}{Na(l)}$$

A substance which brings about the reduction of another substance is called a reducing agent. There are several common reducing agents:

Hydrogen	H_2
Carbon	C
Carbon monoxide	CO
Sulphur dioxide	SO_2
Hydrogen sulphide	H_2S

Compare **Oxidation**.

Refining A process which involves either the removal of impurities from a substance or the extraction of a substance from an impure **mixture**.

(a) **Metals**. When a metal is first extracted from its **ores** it is often not in a totally pure state. It may be acceptable for some uses but not for others. There may be several stages in the refining of a metal to get it into a very pure state.

(b) **Petroleum**. This is a complex mixture of **hydrocarbons**. The first stage of refining involves separating this mixture into different fractions on the basis of **boiling point**. The process is called *fractional distillation*. Each fraction is still a mixture, though not so complicated as before. The way in which the fractions are treated depends on the type of refinery and what products are required. The major product might be petrol or it might be for use in making other chemicals.

Reflection One of three things which can happen to **radiation** in one medium when it meets the surface of a second medium (see **Absorption, Refraction**). Reflected radiation bounces back into the first medium.

The two laws of reflection allow us to predict where a given *incident ray* will go after reflection at a given point, the *point of incidence*. Angles are measured from the *normal*. This is the perpendicular to the surface at the point of incidence.

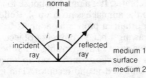

(a) The reflected ray is in the same plane as the incident ray and the normal.

(b) The angle of reflection, r, equals the angle of incidence, i.

The reflection of **water waves** can be shown by experiments using a ripple tank.

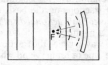

Reflection by a concave barrier

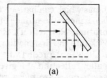 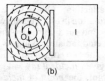

(a)   (b)

Reflection of (a) plane and (b) circular waves at a straight barrier

Reflex action Reflex action occurs in most animals and in vertebrates. It is a rapid response to a **stimulus** over which the animal has no control. Reflex actions can be important in protecting animals from injury; e.g. withdrawing a hand from a hot object.

The *nerve impulse* responsible for reflex actions constitutes a *reflex arc* which is set up when a nerve impulse is initiated in the receptor. The impulse is transmitted along a *sensory neurone* to the **spinal cord** where it crosses a *synapse* to a *motor neurone*. This pathway enables the response to be very rapid.

When a reflex arc operates, nerve impulses are also sent from the spinal cord to the **brain**. Thus, although the response is initiated by the spinal cord, it is through the brain that the animal is aware of what is happened. See **Nervous system**.

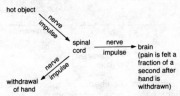

Refraction One of three things which can happen to **radiation** in one medium when it meets the surface of a second (see **Absorption, Reflection**). After refraction the radiation travels on through the second medium almost always in a different direction. If it is not different, it has not been refracted. The two laws of refraction allow us to predict where a given *incident ray* will go after refraction at a point, the *point of incidence*. Angles are measured from the *normal* or *perpendicular* to the surface at the point of incidence.

(a) The *refracted ray* is in the same plane as the incident ray and the normal.

(b) For a given pair of media, the sine of the angle of incidence divided by the sine of the angle of refraction is constant (Snell's law).

This constant is the *refractive index (n)*. It relates to the speeds of the radiation in the two media (see also **Colour, Lens**). When light passes from a less dense medium to a more dense medium it is refracted towards the normal in the more dense medium.

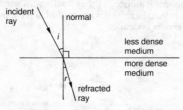

This is caused by the change of speed of the light. When light passes from a more dense medium to a less dense medium the converse is true. When light from a submerged object passes from water to air it is refracted. This causes the object to appear less deep than it really is.

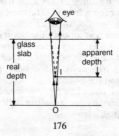

Water waves are refracted when passing from water of one depth to water of a different depth.

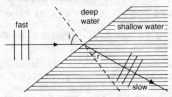

Refraction of water waves

Refrigeration A method for transferring **energy** from a cool box to the warmer outside air. This is the opposite direction to normal net energy transfer and is achieved by a pump powered by a motor. The pump circulates a special fluid around a closed loop.

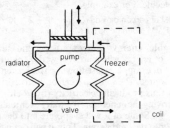

When the pump raises the fluid's **pressure** it becomes a **liquid** and gives off **latent heat**. At this stage the liquid goes through the black radiator coil outside so that heat can be transferred to the air. Further around the loop the liquid is forced through the valve to the low pressure side of the loop. The lower pressure causes the liquid to evaporate absorbing latent heat from the surroundings. At this stage the substance is within the cool box. Devices like this which transfer heat from one place to another are often called *heat pumps*.

Relative atomic mass (RAM) The relative atomic mass of an **element** is the average **mass** of the atoms of the element compared with an atom of the **isotope carbon**—12 which is given the value of exactly 12. See **Relative molecular mass**.

Relative molecular mass (RMM) The **mass** of one molecule of a **compound** calculated by adding together the **relative atomic mass** of each atom within the molecule. For example:

RMM of carbon dioxide (CO_2) = $12 + 2 \times 16 = 44$
RMM of ammonia (NH_3) = $14 + 3 \times 1 = 17$

Renewable energy sources A general term for sources of energy which are renewed by natural processes. For example, a fast-flowing river is a renewable energy source. The water in it can be used to drive **generators** thus producing electricity (**hydroelectricity**). After driving the generators the water eventually flows out to sea. Some of it will evaporate into the air and return onto the land as rain. Thus it will drain into the rivers ready to be used again. Other examples of renewable energy sources are: solar cells, wind turbines, wave generators and geothermal energy units.

At present most of the world's energy is obtained from nonrenewable energy sources (see **Fossil fuels**). Unfortunately these sources will run out within the next hundred years or so. Many scientists believe that renewable energy sources are the only answer to a future energy shortage. There are many research projects underway to develop renewable energy sources for future use. Renewable energy sources also have the advantage that they cause little pollution.

Resistance (R) The opposition of a **circuit** element or section to the flow of **charge (current, I)**. The unit of resistance is the ohm. The current I in a **metal** sample at a constant **temperature** is proportional to the **voltage**, V across its ends: $I \propto V$ *(Ohm's Law)*. We define the sample's resistance R from this. $R = V/I$ (rearranging this equation $V = I \times R$, $I = V/R$). All normal circuit components have resistance. As they oppose current there is energy transfer producing a **temperature** rise (see **Fuses**). The **power, P**, involved is the rate of energy transfer; it is given by:

$$P = V \times I \qquad (= V^2/R \text{ or } I^2 \times R)$$

Temperature often alters a sample's resistance. In most metals, the increase in temperature makes R rise. In semiconductors, carbon and insulators, R falls. *Resistors* may be arranged in *series* or in *parallel* in a circuit.

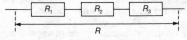

Three resistors in series

179

When arranged in series the total resistance, R, of the group is given by;

$$R = R_1 + R_2 + R_3$$

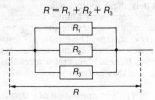

Three resistors in parallel

When arranged in parallel the total resistance, R, of the group is given by:

$$\frac{1}{R} = \frac{1}{R_1} + \frac{1}{R_2} + \frac{1}{R_3}$$

Resonance The large amplitude vibration of a system when it is driven at a **frequency** close to its *natural frequency*. Here are some examples of resonance;
(a) If one sings a note near a piano some of the strings will start to vibrate in sympathy.
(b) The tuning circuit of a radio set will pass high-amplitude signals only if their frequency is near its natural value.
(c) Resonance between the wind and the Tacoma Narrows Bridge (USA) caused the bridge to shake itself apart in 1940.

The resonance tube is a piece of apparatus which can be used to find the **speed of sound** in air. Its length
180

is changed until the air column resonates with a tuning fork of known frequency.

Resources The products which are available within an area. The resources of an area may be described in many different ways. For example:
— the amount of food which can be grown.
— what mineral deposits are present.
— what **energy** sources it has in the form of **coal, oil, natural gas, petroleum**, etc.
See **Resource scarcity**.

Resource scarcity Many **resources** are nonrenewable, they can only be used once. For example, **coal** is a nonrenewable energy source. Once burnt it cannot be used a second time. Much of the **metal** which is obtained from ores is only used once. When the metal article is no longer any use it is thrown away. The continual use of nonrenewable resources results in a resource scarcity.

Respiration The chemical reactions by which organisms release **energy** from food such as **glucose**. Respiration may occur in the presence or in the absence of oxygen.
(a) *Aerobic respiration* occurs in the presence of oxygen within the *mitochondria* of cells.

$$\text{glucose} + \text{oxygen} \rightarrow \text{carbon dioxide} + \text{water}$$
$$C_6H_{12}O_6 + 6O_2 \qquad 6CO_2 \qquad + 6H_2O$$

(b) *Anaerobic respiration* occurs in the absence of oxygen within the cytoplasm of cells. It provides less energy than aerobic respiration.

$$\text{glucose} \xrightarrow{\text{muscle cells}} \text{lactic acid}$$

$$\underset{C_6H_{12}O_6}{\text{glucose}} \longrightarrow \underset{2CH_3CHOHCOOH}{\text{lactic acid}}$$

Fermentation is a form of anaerobic respiration.

$$\underset{C_6H_{12}O_6}{\text{glucose}} \longrightarrow \underset{2C_2H_5OH}{\text{ethanol}} + \underset{2CO_2}{\text{carbon dioxide}}$$

Reversible reaction This is a reaction which can go either way depending on the reaction conditions, e.g.

$$Fe_2O_3(s) + 3H_2(g) \rightleftharpoons 2Fe(s) + 3H_2O(g)$$

Hydrogen can react with iron (III) oxide or steam can react with iron. In both cases an **equilibrium** will exist with all four substances present unless the products are removed. Reversible reactions are shown in equations by using the symbol $\rightleftharpoons$.

Root The part of a flowering plant which normally grows underground in the **soil**. Its functions are:
(a) The absorption of water and **mineral salts** from the soil.
(b) To anchor the plant in the soil.
(c) In some plants, such as the turnip and carrot, it acts as a food store.

Roughage (or fibre) A component of the human **balanced diet**, consisting mainly of **cellulose** from plant cell walls. Although indigestible to humans, roughage plays an important role in digestion. It adds bulk to the food and enables the muscles of the **alimentary canal** to grip the food and keep it moving along by *peristalsis*.

Rust The red-brown product of the **corrosion** of iron or mild steel which has been exposed to water and air. It is hydrated iron (III) oxide ($Fe_2O_3.xH_2O$). Rusting is of great economic importance. Renovation and prevention work on rusting costs British Industry, for example, many millions of pounds each year. There are several commonly used methods of preventing rust.

Method	Example
Covering in grease or oil	cycle chain
Covering in paint	motor car body
Covering in plastic	wire mesh fencing
Electroplating	chrome car bumper
Galvanizing	coal bunker
Sacrificial protection	ship's hull

Saliva Fluid secreted by the salivary glands into the mouths of many animals. Its function is to moisten and lubricate food so that it can be swallowed more easily. In some mammals, including humans, saliva contains the **enzyme** salivary amylase (sometimes called *ptyalin*) which digests **starch** into maltose.

where G = glucose

Salt This is a **compound** formed when the **hydrogen** of an **acid** is totally or partially replaced by a metal. When an acid reacts with a metal the result is a salt and hydrogen gas, e.g.

$$Zn(s) + 2HCl(aq) \rightarrow ZnCl_2(aq) + H_2(g)$$

When an acid reacts with a **base** the result is a salt and water, e.g.

$$NaOH(aq) + HNO_3(aq) \rightarrow NaNO_3(aq) + H_2O(l)$$

Salts may sometimes also be made by the direct combination of two **elements**, e.g.

$$2Na(s) + Cl_2(g) \rightarrow 2NaCl(s)$$

If only one hydrogen **atom** of a *dibasic acid* is replaced the result is an *acid salt*, e.g. sodium hydrogen carbonate $NaHCO_3$. The name of the salt is derived from the metal and the acid used.

sulphuric acid → sulphates
nitric acid → nitrates
hydrochloric acid → chlorides

For example, the name of the salt formed by the reaction of copper (II) oxide and dilute sulphuric acid is copper sulphate.

$$CuO(s) + H_2SO_4(aq) \rightarrow CuSO_4(aq) + H_2O(l)$$

Secondary sexual characteristics These are features which distinguish between adult male and female animals (excluding the reproductive organs). For example, the lion's mane and the antlers on a stag. In humans, secondary sexual characteristics include breast developed in females and facial hair in males. The development of these features is usually controlled by **hormones**. Compare **Primary sexual characteristics**.

Seed This develops from an ovule after **fertilization** in flowering plants. The seeds of a plant are enclosed in a **fruit** (which develops from the ovary).

Section through bean seed

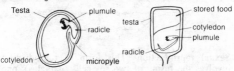

Within the seed, the **embryo** becomes differentiated into an embryonic **shoot** bud (plumule) and **root** (radicle) and either one or two seed leaves (cotyledons). Given the right conditions, **germination** will occur and the seed will grow into a new plant. See **Fruit and seed dispersal**.

Sensitivity The ability of plants and animals to respond to **stimuli**, such as heat, light, sound etc., resulting from changes in their **environment**. Sensitivity makes organisms aware of changes in their environment, thus they can make appropriate responses to any changes which occur. Certain parts of animals, such as the eyes, ears and skin, are specialized in sensing particular environmental stimuli. They are called sense organs or receptors. Similarly, plant **tissues** such as *shoot tips* are receptors and are important in **tropisms**.

Stimuli from the environment cause responses to be initiated in specialized structures called *effectors*. The muscles in our bodies are examples of effectors.

The responses which an organism makes constitutes its *behaviour*. See **Nervous system**.

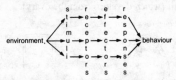

Sexual reproduction This form of reproduction involves the fusing of two sex cells (**gametes**), one from a male parent and the other from a female parent, to form a zygote. The fusing process is called **fertilization**. The gametes are haploid; however, the resulting zygote has a diploid number of **chromosomes**. After ferilization the zygote divides repeatedly, ultimately resulting in a new organism. Thus in humans;

Unlike **asexual reproduction**, the offspring of sexual reproduction are genetically unique (with the exception of identical twins) because they have obtained half of their chromosomes from their male parent and half from their female parent. Thus each fertilization produces a new combination of chromosomes unique to the new organism formed.

Shoot The part of flower which is above the **soil**.

This is often composed of the **stem, leaves**, buds and **flowers**.

Short circuit   A path of low **resistance** between two points in a **circuit**. The large flow of **charge** which results from a short circuit draws a heavy **current** from the source. The **heat** produced by this current should be enough to melt the **fuse** and break the circuit before any damage is done to the wiring of the circuit or the load.

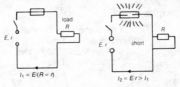

$$I_1 = E/(R = r)$$

$$I_2 = E/r > I_1$$

SI units   Scientific measurements are usually taken in SI units (Système International d'Unités). There are 7 base units including:

length	metre	m
mass	kilogram	kg
time	second	s

Other units are derived from these base units, e.g.

volume	cubic metres	m^3
density	kilograms per cubic metre	kg/m^3
acceleration	metres per second per second	m/s^2

Some derived units are rather complicated and have been given special names, e.g. **pressure**; the derived unit is the kilogram metre per second per second per square metre $(kg m/s^2)/m^2$. This is more normally called the pascal (Pa).

The size of a unit can be altered by using a series of prefixes.

Prefix	Symbol	Meaning
micro	μ	$\times 10^{-6}$ ($\times$ 1/1 000 000)
milli	m	$\times 10^{-3}$ ($\times$ 1/1000)
kilo	k	$\times 10^3$ ($\times$ 1000)
mega	M	$\times 10^6$ ($\times$ 1 000 000)

For example:

$$1 \text{ millimetre} = 10^{-3} \text{ metre}$$
$$1000 \text{ millimetres} = 1 \text{ metre}$$
$$1 \text{ kilometre} = 1000 \text{ metres}$$
$$0.001 \text{ kilometre} = 1 \text{ metre}$$

Skeleton The hard framework of an animal which supports and protects the internal **organs** and gives the animal shape. It also provides a structure for **muscle** attachment and works in concert with muscles to produce movement. In some animals the skeleton lies outside the body and in others the skeleton is contained within the body.

(a) *Exoskeleton* (or external skeleton). A skeleton lying outside the body of some invertebrates. Common examples are the tough *cuticle* of insects and the hard shells of crabs (see illustration opposite). Some organisms have the ability to shed and renew their exoskeletons periodically to allow **growth**. This process is called *moulting* or *ecdysis*.

(b) *Endoskeleton* (or internal skeleton). A skeleton lying within an animal's body. For example, the bony skeleton of vertebrates such as humans. Endoskeletons get bigger as part of the growth process of an animal. The main features of the human endoskeleton are shown overleaf.

Skin A layer of *epithelial cells*, *connective tissue* and associated structures that covers most of the body of vertebrates.

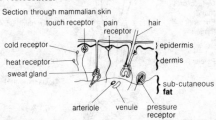

Section through mammalian skin

The skin of mammals can be divided into two main layers.
(a) The outer layer is called *epidermis*. It consists of;
  (i) the *cornified layer*. Dead cells forming a tough protective outer coat.

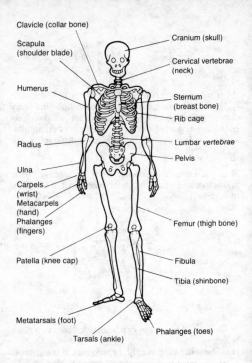

Clavicle (collar bone)

Scapula (shoulder blade)

Cranium (skull)

Cervical vertebrae (neck)

Humerus

Sternum (breast bone)

Rib cage

Radius

Lumbar *vertebrae*

Ulna

Pelvis

Carpels (wrist)

Metacarpels (hand)

Phalanges (fingers)

Femur (thigh bone)

Patella (knee cap)

Fibula

Tibia (shinbone)

Metatarsais (foot)

Phalanges (toes)

Tarsals (ankle)

The human endoskeleton

(ii) the *granular layer*. Living cells which eventually die and form the cornified layer.

(iii) the *Malpigian layer*. Cells which are actively dividing to produce new epidermis.

(b) The inner layer is called the *dermis*. It is a thicker layer than the epidermis and contains **blood capillaries**, hair follicles, sweat glands, receptor cells sensitive to touch, **heat**, cold, pain and **pressure**.

Beneath the dermis there is a layer of **fat** storage cells. These cells act as a food store for the body and also provide heat **insulation**.

The functions of the mammalian skin are:

(a) Protects against injury and the entry of microorganisms which may be harmful to the body.

(b) Reduces water loss by **evaporation**.

(c) Acts as a receptor for certain environmental **stimuli**.

(d) In homiothermic animals (animals whose body temperatures remain fairly constant despite environmental conditions) it is important in body **temperature regulation**.

Smell The ability of animals to detect odours. In humans the nose is the **organ** of smell. The receptor cells involved are in the nasal cavity. They are sensitive to chemical **stimuli**. See **Sensitivity**.

Sodium A soft grey metallic element from **group** 1 of the **periodic table**. The metal is easily cut with a knife to reveal a silvery surface which rapidly tarnishes on exposure to air. Sodium reacts vigorously with cold water.

$$2Na(s) \ + \ 2H_2O(l) \ \rightarrow \ 2NaOH(aq) \ + \ H_2(g)$$

sodium	water	sodium	hydrogen
		hydroxide	

Sodium is stored under oil because of its reactivity to air and water.

Sodium metal is obtained by the **electrolysis** of molten sodium chloride. It is used as a coolant in *fast-breeder nuclear reactors* and in the manufacture of the petrol additive tetramethyl lead (see **Lead**).

Sodium **ions** (Na^+) are an important constituent of the fluids in animal **tissues**. One way in which humans obtain sufficient sodium is to add salt (sodium chloride) to food either during cooking or when eating; however, excessive amounts of sodium cause damage to the body. There are several compounds of sodium which figure prominently in our everyday lives.

(a) *Sodium carbonate* (Na_2CO_3). This compound is made in the *Solvay Process*. It is important in the manufacture of glass. Unlike most carbonates, sodium carbonate does not decompose when heated. Its aqueous solution is alkaline. It may be used to remove permanent hardness from water (see **Hardness of water**).

(b) *Sodium chloride* (NaCl). Known as common salt, and obtained from rock salt. As well as its use for flavouring and preserving foods, it is used in the production of sodium metal and sodium hydroxide.

(c) *Sodium hydrogencarbonate* ($NaHCO_3$). Commonly called baking powder. It is decomposed either by heat or by the action of **acids**.

$$\overset{\text{heat}}{2NaHCO_3(s) \rightarrow Na_2CO_3(s) + CO_2(g) + H_2O}$$

sodium   sodium   carbon   water
hydrogencarbonate   carbonate

$$NaHCO_3(s) + HCl(aq) \rightarrow NaCl(aq) + H_2O(l) + CO_2(g)$$

sodium   **hydrochloric**   sodium   water   **carbon**
hydrogen-   **acid**   chloride   **dioxide**
carbonate

Sodium hydrogencarbonate is also used in fire extinguishers and anti-indigestion powders.

(d) *Sodium hydroxide* (NaOH). This is made by the electrolysis of brine. It is a caustic **alkali** (it will corrode and burn organic material such as flesh) and solutions have **pH** > 10. It has many uses in industry, e.g. making soaps and paper.

(e) *Sodium nitrate* ($NaNO_3$). This is used as a **fertilizer** and in the preservation of meat.

(f) *Sodium sulphate* (Na_2SO_4). The hydrated form of this salt (.$10H_2O$) is commonly known as Glauber's Salt. The sulphate is used in the manufacture of paper.

(g) *Sodium thiosulphate* ($Na_2S_2O_3$). This compound is used in the photographic process. It is used to *fix* the negative and is often called *hypo*. It reacts with unreacted silver bromide, thus ensuring that no further reaction with light occurs.

Soil A typical sample of soil contains the following components.

(a) Inorganic particles — the result of weathering of rocks.

(b) Water.

(c) **Humus** — organic material.
(d) Air.
(e) **Mineral salts**.
(f) Microorganisms.
(g) Other larger organisms such as earthworms.
Soil is important for several reasons.
(a) It provides a habitat for a wide range of organisms.
(b) It provides plants with a medium for growth and supplies them with water and mineral salts.
(c) The decomposition of dead organisms in soil releases minerals which can be reused by other living organisms.

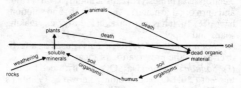

Soil is often classified into three distinct types.
(a) *Sandy (light) soil.* This contains a high proportion of larger inorganic particles and hence large air spaces between them. Thus sandy soils are well aerated and have good drainage. Unfortunately good drainage also tends to allow leaching (washing out) of mineral salts. See diagram opposite.
(b) *Clay (heavy) soil.* This has a high proportion of small particles and hence small air spaces. It retains mineral salts and water but is poorly aerated and may easily become waterlogged.

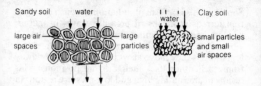

Sandy soil   water   water   Clay soil
large air   large   small particles
spaces   particles and small
large   air spaces

(c) *Loam soil.* This consists of a balance of particle types and a good humus content. The soil is well aerated and drains well whilst retaining water and mineral salts. Loam is the most fertile soil.

Solid A solid is a substance whose **atoms** or **molecules** are fixed in positions and do not have the freedom of movement found in a **liquid** or a **gas**. Atoms and molecules are held in a lattice by **bonds**. It is only when these bonds are broken that the atoms and molecules are able to move and the solid *melts*.

Solubility The solubility of a *solute* in a *solvent* is the extent to which it dissolves. In general the solubility of a particular solute increases as the **temperature** of the solvent increases.

In a polar solvent, an **ionic compound** will have a higher solubility than a **covalent compound**. For example, in water, sodium chloride is much more soluble than **methane**. Solubility is usually measured in grams of solute per 100 g of solvent at a stated temperature. However other units such as mole/dm^3 and mole/100 g are also used.

Sound This is a form of **radiation** in which energy is transferred by means of **pressure waves** in matter, hence sound waves cannot travel through a **vacuum**. A vibrating source pushes particles of matter closer together; as they move apart they move further away from each other than their original positions. This results in alternating high and low pressure regions in the matter. A plot of pressure in the medium at different points at a given time produces a sine wave. A similar pattern is produced by plotting the pressure at a point as time passes.

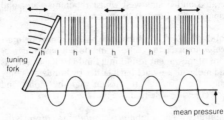

h — high pressure (compression)
l — low pressure (or rarefaction)

Sound is thus a wave and shows all wave properties; **absorption, reflection, refraction, diffraction** and *interference*. However sound is a *longitudinal* wave and cannot be polarized.

Like other waves, pressure waves have a **frequency spectrum**. The approximate range of sound which can be detected by the human ear is 20–20 000 Hz. Radiation from the region below 20 Hz is known

196

as *infrasound* and from the region above 20 000 Hz is known as **ultrasonics**. See **Speed of sound**.

Species A unit used in the **classification** of living organisms. It describes any group of organisms which share the same general physical characteristics and can *mate and produce fertile offspring*. For example, all dogs, despite a wide variation in shape and size, are all of the same species.

Spectrum A type of graph which shows how different types of **radiation** relate to wavelength (and **energy**). For example, **electromagnetic waves** appear on the electromagnetic spectrum. Visible light is a part of this spectrum. See **Colour**.

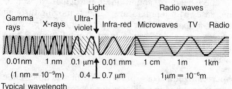

Typical wavelength

Speed (c) The rate of change of distance. The unit of distance is the metre per second, m/s. Speed is a scalar measure since the direction of movement is not important. This makes it different from velocity which is a *vector* quantity.

Speed of light (or **velocity of light**) The speed of light in empty space, 300 000 km/s. In space all **electromagnetic waves** travel at this speed. In matter they move more slowly at speeds which depend on the nature of the substance and the wavelength. See **Refraction**.

Speed of sound The speed of sound travelling through matter depends on the elasticity of the medium and its **density**. The density and elasticity of **solids** and **liquids** are far higher than those of **gases** hence the speed of sound is lowest in gases. Here are some values for the speed of sound in different media.

Medium	Speed m/s
Air	330
Water	1500
Steel	6000

The speed of sound in a particular substance increases with **temperature**. Objects moving through air at a speed greater than sound have *supersonic* speeds and shock waves form around them. These cause Concorde's *sonic boom* and the crack of a long whip.

Spermatozoon (or **sperm**) A microscopic male **gamete** formed in animal **testes**. A sperm usually has a *flagellum* and is able to move. Sperms are released from the male in order to fertilize the female gamete. See **Fertilization**.

Spinal cord The part of the central nervous system of a vertebrate which is enclosed within and protected by the backbone. It is a cylindrical mass of *nerve cells* which connects the **brain** to the other parts of the body via the spinal nerves. There are three distinct regions in the spinal cord.

(a) An inner layer of grey matter which consists of nerve cell bodies.
(b) An outer layer of white matter which consists of nerve fibres running the length of the cord.
(c) A fluid-filled central canal.

section through spinal cord

The spinal cord conducts *nerve impulses* to and from the brain and is also involved in **reflex actions**.

Spleen An organ in the abdomen. In most vertebrates it is found near the stomach. It has several important functions:
(a) It produces **white blood cells**.
(b) It destroys worn out **red blood cells**.
(c) It filters foreign bodies from the blood.

Spore A (usually) microscopic reproductive unit consisting of one or several **cells** which have become detached from the parent organism and will ultimately become a new individual. Spores are involved in both **asexual** and **sexual reproduction** (as **gametes**). They are produced by certain plants, fungi, bacteria and protozoa. Some spores form a resistant resting stage during a **life cycle** while others allow a rapid colonization of new habitats.

Spore release in the bread mould *Mucor*

sporangium
spores
spores being released
hypha
mycelium

Stamen The male part of the **flower**. Each stamen consists of a stalk (filament) and at the end of each filament there is an anther. Within the anthers there are pollen grains containing the male **gametes**. See **Fertilization in plants**.

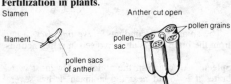

Stamen

filament

pollen sacs of anther

Anther cut open

pollen grains

pollen sac

Standard temperature and pressure (STP) In science, many of the measurements we might take of the sample vary with **temperature** and/or **pressure**. To avoid confusion it is useful to quote values for a specific temperature and pressure. Thus, if the sample is kept under these standard conditions in a laboratory anywhere in the world, the measurements obtained should be the same. The standard conditions chosen are 0°C (the **melting point** of pure ice) and 101 325 Pa (air pressure of 1 atmosphere).

Starch A *polysaccharide* **carbohydrate** which consists of chains of **glucose** units. Starch is important as an **energy** store in plants and is synthesized by plants during **photosynthesis**. It is readily converted into glucose by *amylase* **enzymes**.

States of matter   Substances can exist in three states of matter; **solid, liquid** and **gas**. The state in which a substance exists depends upon its **temperature** and the **pressure** exerted on it. Not all substances exist in all states. The letters (s) = solid, (l) = liquid, (g) = gas, (and (aq) = aqueous solution) are sometimes used in chemical **equations** to show the states of the reactants and products, e.g.

$$2Na(s) + 2H_2O(l) \rightarrow 2NaOH(aq) + H_2(g)$$

Static electricity   This is a study of **charges** and how they affect each other. Static often appears on an **insulator** when it is rubbed and the charge may last for a long time. The discharge of static which has built up on a gramophone record can be heard as a 'crackle'. The discharge of static is a potential hazard in areas such as mines, factories and aircraft fuelling systems. A spark in the presence of inflammable material in the air may result in an explosion.

Steel   Steels are **alloys** which contain **iron** as the main constituent. Other elements presnt will determine the properties of the steel. The two best known types of steel are *mild steel* and *stainless steel*.

(a) Mild steel contains iron with small amounts of **carbon**. It is used for car bodies and household goods such as freezers and cookers. It is cheap to make, however, it **rusts** easily in the presence of moisture and oxygen and so it must be protected. Common methods of protection are painting, greasing, enamelling, *galvanizing* and coating in plastic.

(b) Stainless steel contains iron, carbon, chromium and nickel. Common uses of stainless steel are cutlery, surgical instruments and sink tops. It is not corroded by oxygen, however its use is limited because it is much more expensive to produce than mild steel.

The iron made in a **blast furnace** contains impurities which make it brittle, e.g. carbon (more than is needed for mild steel). This iron is made into steel by blowing oxygen through the molten iron and thus oxidizing the impurities which are given off as gases. When this is done other elements are added to the melt to produce the steel required.

Stem The part of a flowering plant which bears the buds, **leaves** and **flowers** (see diagram opposite). The functions of the stem are:

(a) to transport water, **mineral salts** and **carbohydrate**.

(b) to raise the leaves above the soil so they get the maximum amount of air and light.

(c) to raise flowers, thus aiding the process of **pollination**.

(d) in green stems, **photosynthesis**.

Stimulus Any change in the **environment** of an organism which may produce a response in the organism. For example, a change in **temperature** or a change in **pressure**. See **Sensitivity**.

Stomach This is a muscular sac in the anterior region

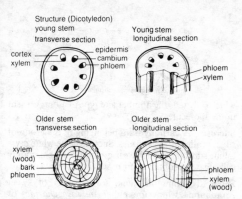

Structure (Dicotyledon)
young stem
transverse section

cortex
xylem
epidermis
cambium
phloem

Young stem
longitudinal section

phloem
xylem

Older stem
transverse section

xylem
(wood)
bark
phloem

Older stem
longitudinal section

phloem
xylem
(wood)

of the **alimentary canal**. In vertebrates food passes into the stomach from the mouth via the oesophagus.

In the stomach, food is mechanically churned by the peristaltic action of the muscular walls and the **digestion** of **protein** is started. In herbivores the stomach has several chambers. This structure is designed to assist with the digestion of **cellulose**.

The amount of time which food spends in the stomach depends on its nature. From the stomach it passes into the small intestine through a ring of muscle called the *pyloric sphincter*.

Structural formula The structural formula of a substance shows the bonds between its atoms and their positions relative to each other. The structural formula gives more information than the **molecular formula**,

ethane
C_2H_6

$$
\begin{array}{ccc}
 & H & H \\
 & | & | \\
H - & C - C & - H \\
 & | & | \\
 & H & H
\end{array}
$$

ethanol
C_2H_5OH

$$
\begin{array}{cccc}
 & H & H \\
 & | & | \\
H - & C - C & - O - H \\
 & | & | \\
 & H & H
\end{array}
$$

molecular formula   structural formula

which only tells the number and type of atoms present in the molecule.

Sucrose A **carbohydrate** belonging to the group of **sugars** called *disaccharides*. Sucrose is the white crystalline substance which we use at home and commonly call sugar. It is obtained from sugar beet or sugar cane.

Sugar A common name for a series of sweet **compounds** which are either *monosaccharides* or *disaccharides*. For example;
(a) monosaccharides — **glucose**, fructose, ribose.
(b) disaccharides — maltose, **sucrose**.
See **Carbohydrate**.

Sulphur (S_8) A yellow nonmetallic **element** which may exist in two forms or *allotropes*. *Rhombic sulphur* is the stable form below 96 °C and *monoclinic sulphur* is stable above that **temperature**.

Sulphur belongs to **group 6** of the **periodic table** and is reactive to both **metals** and **nonmetals**. It occurs in nature both uncombined and as metal suphides such as galena (PbS) and pyrite (FeS). Sulphur and sulphur-containing compounds are also found in **petroleum, natural gas** and **coal**. Such impurities are a problem as the **combustion** of any fuel containing them will produce sulphur dioxide, one of the main causes of **acid rain**. In modern processes sulphur and sulphur-containing compounds are either removed from fuels before combustion, or any sulphur dioxide gas produced is removed from the exhaust gases. Sulphur is an important constituent of some drugs, such as *sulphonamides*, and is used to *vulcanize* rubber; however, its main use is in the manufacture of **sulphuric acid** in the *contact process*.

Sulphuric acid (H_2SO_4) A very important chemical used in many processes. It is a colourless, oily liquid which is a strong **acid** and a vigorous oxidizing agent (see **Oxidation**). It is made from *sulphur dioxide* in the *Contact Process*.

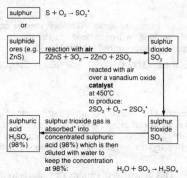

sulphur	$S + O_2 \rightarrow SO_2^*$

or

| sulphide ores (e.g. ZnS) | reaction with **air**
$2ZnS + 3O_2 \rightarrow 2ZnO + 2SO_2$ | sulphur dioxide SO_2 |

reacted with air over a vanadium oxide **catalyst** at 450°C to produce: $2SO_2 + O_2 \rightarrow 2SO_3^*$

| sulphuric acid H_2SO_4 (98%) | sulphur trioxide gas is absorbed* into concentrated sulphuric acid (98%) which is then diluted with water to keep the concentration at 98%: | sulphur trioxide SO_3 |

$H_2O + SO_3 \rightarrow H_2SO_4$

Sulphuric acid reacts chemically in several ways.

205

(a) **As an acid.** Dilute sulphuric acid reacts with **metals, bases** and *carbonates* to form *sulphates*.

$$Mg(s) + H_2SO_4(aq) \rightarrow MgSO_4(aq) + H_2(g)$$
$$CuO(s) + H_2SO_4(aq) \rightarrow CuSO_4(aq) + H_2O(l)$$
$$ZnCO_3(s) + H_2SO_4(aq) \rightarrow ZnSO_4(aq) + H_2O(l) + CO_2(g)$$

Concentrated sulphuric acid reacts with chlorides and nitrates to form hydrogen chloride and nitric acid respectively.

$$H_2SO_4(l) + NaCl \rightarrow NaHSO_4(s) + HCl(g)$$
$$H_2SO_4(l) + NaNO_3 \rightarrow NaHSO_4(s) + HNO_3(g)$$

(b) **As a dehydrating agent.** Concentrated sulphuric acid is sometimes used to dry gases. It can also be used to remove the atoms which make water from substances.

(c) **As an *oxidizing agent*.** Although **copper** cannot directly replace hydrogen from acids, the metal is oxidized by concentrated sulphuric acid.

$$Cu(s) + 2H_2SO_4(l) \rightarrow CuSO_4(s) + SO_2(g) + 2H_2O(l)$$
anhydrous
copper (II)
sulphate

The reaction between concentrated sulphuric acid and water is very **exothermic**. When diluting concentrated sulphuric acid, it is important always to add the acid to the water (by stirring) and NOT the other way around.

Taste This describes the ability of animals to detect flavours. In humans the receptor cells for taste are called taste buds. They are sensitive to certain chemical **stimuli**. They are restricted to the mouth, particularly on the tongue. There are four types of taste bud sensitive to sweet, sour, salt and bitter.

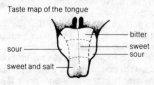

Taste map of the tongue

- bitter
- sweet
- sour

sour

sweet and salt

The senses of taste and **smell** often work together to identify different types of food. See **Sensitivity**.

Teeth Structures within the mouth of vertebrates which are used for biting, tearing and crushing food before it is swallowed.

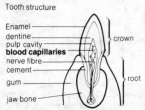

Tooth structure

Enamel
dentine
pulp cavity
blood capillaries
nerve fibre
cement
gum
jaw bone

crown

root

Enamel. This covers the exposed surface of the tooth (the crown). It contains calcium phosphate and is the hardest substance in a vertebrate's body. It is well suited to biting food without itself being damaged.

Dentine. This substance is similar to **bone** and forms the inner part of the tooth.

Pulp. The soft **tissue** in the centre of the tooth. It contains the **blood capillaries**, which supply food and **oxygen**, and the nerve fibres which register pain if the tooth is damaged.

Root. The part of the tooth within the gum and embedded in the jawbone. The outer surface of the root is covered in a substance called cement. A series of fibres hold the tooth in place. One end of them is embedded in the cement and the other in the jawbone.

Most human beings are *omnivores.* The adult human jaw contains four types of teeth.

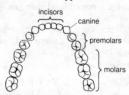

The arrangement of teeth in an adult human jaw (bottom and top jaws are the same)

Incisors. Chisel-shaped teeth at the from of the mouth used for biting off pieces of food.

Canines. Pointed teeth at each side of the incisors. They are used for ripping off pieces of food. In wild animals which are *carnivores,* such as lions and tigers, these teeth are often very large and also used for killing prey. In *herbivores* these teeth are often small or missing completely.

Premolars. Grinding teeth found between the canines and the molars.

Molars. Together with premolars these teeth are often called *cheek teeth* because of the position they occupy in the mouth. They are broad crowned grinding teeth at the back of the mouth. They crush and grind food before it is swallowed. They are found in both herbivores and omnivores but in carnivores they are replaced by *carnassial* teeth.

See **Dental formula**.

Telescope   An optical instrument which produces enlarged images of distant objects. There are several different types of telescope in common use.

(a) A Galilean telescope (as in opera glasses) combines a converging and a diverging **lens**.

(b) A standard refracting telescope has two converging lenses. This produces more magnification than a Galilean type but the tube needs to be much longer and the image produced is upside down.

(c) A reflecting telescope has a converging mirror in place of an object lens.

(d) A standard radio telescope has the same design as a reflecting telescope but it is designed to detect **radio waves**.

Temperature (T)   The degree of hotness or **heat** level of a substance. It relates to the mean **energy** of the particles of the sample. There are several temperature scales used to show the degree of temperature. In daily life we use *Celsius (centigrade)* or *Fahrenheit*. In science the *absolute* scale is also commonly used.

Scale	Unit	Melting point of pure ice	Boiling point of pure water
Celsius	degree C, °C	0 °C	100 °C
Fahrenheit	degree F, °F	32 °F	212 °F
Absolute	kelvin K	273K	373K

Temperature is measured using a **thermometer**.

Temperature regulation *Homoiothermic* animals maintain their body temperature within a narrow range despite the temperature of their environment. For example, the temperature of a healthy human being is always around 37 °C. This is essential so that the normal reactions of **metabolism** can take place. Here are some ways in which birds and mammals regulate their temperature.
(a) Fat under the skin (subcutaneous) acts as an insulator.
(b) Hair in mammals and feathers in birds trap air which is a good insulator.
(c) In mammals the evaporation of sweat from the surface of the skin has a cooling effect.
(d) Blood vessels near the surface of the skin constrict in response to cold. This diverts blood away from the skin surface so less heat is lost *(Vasoconstriction)*.
(e) Blood vessels near the surface of the skin dilate in response to heat. This brings blood up to the skin surface so more heat is lost to the atmosphere *(Vasodilation)*.

Testis The **sperm**-producing reproductive organ of male animals. Vertebrates such as male humans have a pair of testes which, in addition to sperms, also produce **hormones**. See **Fertilization in humans**.

Thermometer A device used to measure **temperature** or hotness. This relates to the mean **energy** of the particles of the sample. Thermometers work by measuring something which varies with temperature, e.g.
(a) a column of **liquid** in a glass tube; as the liquid gets hotter it expands and the level rises up the tube.
(b) the **voltage** between two metals.

Other properties, such as the pressure of a gas, the **resistance** of a wire, or the colour of a hot surface, may also be used. In each case the device is used at certain known temperatures (often the **boiling** and **freezing points** of water and is then marked in degrees according to the scale used.

Thermostat A device which is used to keep the **temperature** in a place within a particular range. Thermostats are present in a number of common household devices such as cookers, refrigerators, irons, freezers and heating boilers. Many thermostats use a **bimetallic strip**.

Three-pin plug The standard method of joining a cable to a mains socket. A mains cable has three wires.

Wire	Colour of plastic coating	Purpose
live	brown	carries current
neutral	blue	returns current
earth	green/yellow	safety

Each wire goes to a particular pin in the plug.

211

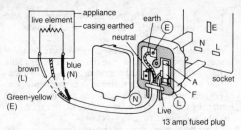

13 amp fused plug

The size of the **fuse** fitted in the plug depends on the **power** of the device to which it is connected.

Tissue In multicellular organisms tissue is a group of similar **cells** which are specialized to carry out a specific function within the organism. For example, **muscle** in animals and *xylem* in plants. The human body contains many different types of tissue.

Transformer A device normally used to transfer electrical **energy** with a change in **voltage**.

An **alternating current** input to the primary coil causes an alternating magnetic field in the core of the transformer. In turn, this changing field induces an alternating current at the secondary (output) coil. The ratio of the output voltage to the input voltage equals the ratio of the number of turns of wire in the output and input coils, i.e.;

$$\frac{V_2}{V_1} = \frac{N_2}{N_1}$$

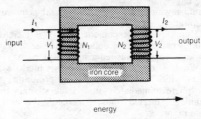

energy

For a *step down* transformer, $V_2 < V_1$ hence $N_2 < N_1$, and for a *step up* transformer the reverse applies $V_2 > V_1$ hence $N_2 > N_1$. The electrical power in the coil of a transformer is calculated by multiplying the current passing through the coil by the voltage across it. The **efficiency** of a transformer is often measured as the ratio of useful output power to total input power.

$$\text{efficiency} = \frac{\text{useful output power}}{\text{total input power}}$$

No energy transfer device has an efficiency of 100%; however, many transformers are very close to this.

Transistor A semiconductor device whose output **current** depends on signals to the base. Its main uses are as an *amplifier* and an *oscillator*.

Transpiration The **evaporation** of water vapour from plant leaves via tiny pores called *stomata*. The rate at which transpiration occurs depends on several environmental factors.

(a) **Temperature** – increased temperature increases

the rate at which water evaporates and thus increases transpiration.

(b) Humidity (water content of air) — increased humidity causes the atmosphere to become saturated with water vapour thus reducing transpiration.

(c) Wind — increased air movements increase the rate at which water evaporates by preventing the atmosphere immediately around the stomata from becoming saturated; thus it increases the rate of transpiration.

It follows from the above that the transpiration rate will be greatest in warm, dry windy conditions. If the rate of water loss by transpiration exceeds the rate of water uptake through the **roots** the plant may begin to wilt.

Tropisms In plant **growth**, movement in response to a **stimulus** such as **light**. These movements are related to the direction of the stimulus, the plant organ growing either towards or away from it. Tropisms are named by adding a prefix which refers to the stimulus involved.

(a) *Geotropism* — a response to gravity.

(b) *Phototropism* — a response to light.

(c) *Chemotropism* — a response to chemicals.

(d) *Hydrotropism* — a response to water.

Tropisms can be either positive or negative depending on whether the response is in the same direction or the opposite direction to the stimulus. They are important because they allow plants to grow in such a way that they can get the maximum benefit from their **environment** in terms of water, light, etc.

Tropisms are caused by a plant **hormone** or *auxin* which accelerates growth by stimulating cell division and elongation. Uneven distribution of auxins causes uneven growth which eventually leads to bending.

Ultrasonic This word describes **sound waves** of higher **frequency** than can be detected by the human ear (about 20 kHz). Some other animals are able to hear some of the ultrasonic range. For example, the *sonar* system of bats typically works at 50 kHz or higher. Ultrasound of megahertz frequencies has many uses in modern life. For example, detecting explosive mines at sea, detecting babies in the womb (typical value 2.25 MHz), and cleaning engine parts.

Ultraviolet A region in the **spectrum** of **electromagnetic waves**. The approximate **wavelength** range is $10^{-10} - 10^{-7}$ m and the approximate **frequency** range is $10^{15} - 10^{18}$ Hz. Ultraviolet **radiation** is produced by white-hot objects and certain gas discharges. It affects photographic film and causes *fluorescence* and *photoelectric effects*. Human skin makes **vitamin D** with this radiation and light skins are tanned by it. However, over-exposure is harmful to both the eyes and skin. The **ozone layer** limits the amount of ultraviolet radiation which reaches the earth's surface.

Unicellular (of an organism) This word describes an organism which consists of only one cell. Unicellular organisms include protozoans, **bacteria** and some algae. Compare **Multicellular**.

Uranium This **element** is a metal which has 92 **protons**. Its atoms are the most massive of all natural elements. Uranium has three **isotopes**; uranium-234, uranium-235 and uranium-238. All three are **radioactive**, being sources of **alpha particles**.

The **nuclei** of uranium-235 atoms can undergo nuclear **fission** when they absorb **neutrons**. The process of nuclear fission itself produces more neutrons. Under certain conditions, a **chain reaction** may take place. This is the basis of **nuclear power** and the atomic bomb.

Urea The main nitrogenous excretory product of mammals. Urea is made in the liver as a product of the *deamination* of excess **amino acids**. It travels from the liver to the **kidney** via the bloodstream and is excreted in the **urine**.

$$H_2N-\underset{\underset{O}{\parallel}}{C}-NH_2 \quad \text{urea}$$

Urine An aqueous solution of **urea** and salts produced by the kidneys of mammals. It is stored in the bladder before being discharged via the urethra.

Uterus (or **womb**) A muscular cavity found in most female mammals. The uterus contains the **embryo** during its development. It receives **ova** from the oviduct and connects to the exterior of the body via the vagina. See **Fertilization in humans, Menstrual cycle, Pregnancy**.

Vacuum A space from which the air has been removed, thus containing very few particles of matter. A cubic metre of air at standard pressure contains around 10^{25} particles. This figure drops to around 10^{14} for a good vacuum on earth. Between the earth and the moon the value is far lower, and between galaxies there may be a true vacuum (containing no particles). Nevertheless the vacuum of outer space contains **energy** in other forms (than matter) such as **force** fields, **radiation** and very low **mass** particles called neutrinos.

Valency This can be thought of as the number of bonds which an atom forms with other atoms. More precisely, the valency of an **element** is the number of **electrons** that it needs to form a **compound** or radical. The electrons may be lost, gained or shared with another atom.

Some elements always have the same valency. Sodium always has a valency of 1. It gives one electron away when it forms the sodium **ion** Na^+ in compounds like sodium chloride NaCl. Oxygen always has a valency of 2. It accepts two electrons when it forms the oxide ion O^{2-} or it forms **covalent compounds** such as CO_2 and SO_2. *Transition elements* have more than one valency, e.g. cobalt = 2 or 3; copper = 1 or 2; iron = 2 or 3. The table overleaf shows the valencies of some common elements and the ions they form.

Variation The difference in characteristics between two members of the same **species**. There are two main types of variation.

(a) *Discrete variation.* In this type of variation only particular values within a given range are possible. For

217

Positive values

+1		+2		+3	
Lithium	Li^+	Calcium	Ca^{2+}	Aluminium	Al^{3+}
Sodium	Na^+	Magnesium	Mg^{2+}	Iron (III)	Fe^{3+}
Potassium	K^+	Zinc	Zn^{2+}		
Silver	Ag^+	Iron (II)	Fe^{2+}		
Ammonium	NH_4^+	Lead	Pb^{2+}		
Hydrogen	H^+	Copper (II)	Cu^{2+}		

Negative values

-1		-2		-3	
Fluoride	F^-	Sulphide	S^{2-}	Phosphate	PO_4^{3-}
Chloride	Cl^-	Oxide	O^{2-}		
Bromide	Br^-	Carbonate	CO_3^-		
Iodide	I^-	Sulphate	SO_4^-		
Hydroxide	OH^-				
Nitrate	NO_3^-				

example, a person's blood may belong to group A, B, AB or O but not somewhere between two groups, i.e. there are no possible intermediate forms. Discrete data does not show a *normal distribution*.

(b) *Continuous variation.* In this type of variation the variable may take any value within a particular range. For example, if a person is between 1.90 and 1.91 m tall they may be 1.900, 1.901, 1.902 m etc. The only limitation is the accuracy with which we measure the height. Another example in humans is **mass**. Continuous variation throughout a population shows a normal distribution about an average value called the mean. (We often represent continuous data as if it were discrete data by placing it in groups.)

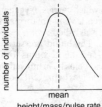

number of individuals

mean

height/mass/pulse rate

Variation within a **species** results from either inherited or environmental factors or a combination of both. For example, a human being inherits **genes** influencing height but is also subject to environmental factors such as nutrition. Inherited variations are considered to be the basis of **evolution** by **natural selection**.

Vegetative reproduction (or **propagation**) **Asexual reproduction** in plants by an outgrowth of some kind from the parent plant. The outgrowth may occur in several different ways. Here are some examples:

Form of outgrowth	Example of plant
bulb	daffodil, onion
corm	crocus, gladiolus
rhizome	iris
stolon	strawberry
tuber	potato, dahlia

Vein A **blood** vessel which transports blood from the **tissues** of the body to the **heart**. In mammals, veins carry deoxygenated blood, with the exception of the pulmonary vein which carries oxygenated blood from the lungs to the heart. Veins are formed from groups of smaller blood vessels called *venules* which carry blood from **capillaries**. The blood pressure in veins is less than in **arteries**, and this accounts for the differences in their structures. Veins have much thinner walls than arteries and have valves to prevent blood flowing away from the heart.

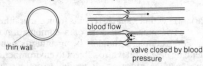

Section through a vein   Valve operation in veins

thin wall

blood flow

valve closed by blood pressure

The largest veins in humans are the *superior vena cava*, which carries blood from the head, neck and upper arms to the heart, and the *inferior vena cava* which carries blood from the rest of the body to the heart.

Velocity (v) An object's velocity is its displacement in unit time. The unit used is the metre per second, m/s. Velocity is a *vector* (compare **speed**). To find velocity we divide the displacement, s, by the time taken, t; $v = s/t$. The slope at any point on a graph of displacement against time gives the velocity at that instant.

Vertebral column (or **backbone**) A series of interlocking bones (called *vertebrae*) and/or *cartilages* which runs along the back from the skull to the tail in vertebrates. It is the principal longitudinal supporting
220

structure of the body. It also protects the **spinal cord** which runs down its centre. See **Skeleton**.

Virus Viruses are the smallest known living organisms, having diameters between 0.025 and 0.25 microns. They are **parasites** infecting animals, plants and **bacteria**. A virus particle consists of a **protein** coat surrounding a length of nucleic acid, either **DNA** or **RNA**.

Virus infecting a bacterium

| bacterium virus becomes attached to bacterium | the virus nuclear material is injected into the bacterium and causes the assembly of new virus parts | the bacterium cell wall is ruptured, releasing many new viruses |

Virus infections in humans include measles, polio and influenza.

Vitamins These are organic **compounds** which are required in small amounts by living organisms. Like **enzymes**, vitamins play a vital role in chemical reactions within the body and often regulate an enzyme's action. If the human diet contains insufficient amounts of vitamins this will result in deficiency diseases. See table overleaf.

Voltage (or **potential difference**) The unit is the volt, V. **Electricity** will only flow between two places which have a potential difference between them. One volt is the potential difference between two points

Vitamin	Rich sources	Effects of deficiency
A	Milk, liver, butter, fresh vegetables.	night-blindness and in children retarded growth.
B_1	Yeast, liver.	Beri-beri; loss of appetite and weakness.
B_2	Yeast, milk.	Pellagra; skin infections, weakness, mental illness.
C	Citrus fruit, fresh vegetables.	Scurvy; bleeding gums, skin disorders.
D	Eggs, cod liver oil.	Rickets; abnormal bone formation.
E	Fresh green vegetables, milk.	Thought to affect reproductive ability.
K	Fresh vegetables.	Blood clotting impaired.

which allows 1 **joule** of **energy** to be obtained when 1 *coulomb* of **charge** moves between the points.

$$\text{volts} = \text{joules/coulomb}$$

Volume The amount of space taken up by an object. The **SI** unit of volume is the cubic metre, m^3.

The volumes of **solids** and **liquids** are fairly constant. The volume of a **gas** depends greatly on the **temperature** and **pressure**. See diagram opposite.

Water (H_2O) Water is an oxide of **hydrogen**.

$$2H_2(g) + O_2(g) \rightarrow 2H_2O(l)$$

volume
of cuboid = a × b × c

volume
of cylinder = $\pi r^2 h$

volume
of sphere = $\frac{4}{3}\pi r^3$

volume
of cone = $\frac{1}{3}\pi r^2 h$

Water is one of the most common **compounds** on the earth, and is essential for all living things. It is a colourless liquid. Here are some of its more important properties.

Freezing point	0 °C
Boiling point	100 °C
Density	1.0 g/cm³

(the maximum density of water occurs at 4 °C)

Water is an atypical liquid in that it expands on solidification, thus ice is less dense than water and floats on it. This unusual expansion is the cause of water pipes bursting in the winter. Pure water does not conduct electricity; however, it can be electrolysed if small amounts of **acid** (H_2SO_4) or **alkali** (NaOH) are added. The products of this **electrolysis** are hydrogen and **oxygen**.

$$2H_2O(l) \rightarrow 2H_2(g) + O_2(g)$$

Two simple tests for the presence (but not the purity) of water are by changing the colour of **anhydrous salts**.

$$CuSO_4(s) + 5H_2O \rightarrow CuSO_4.5H_2O(s)$$
white   blue
$$CoCl_2(s) + 6H_2O \rightarrow CoCl_2.6H_2O(s)$$
blue   pink

One way to test the purity of water is to measure its **boiling point**.

As well as in animals and plants, water is found in the **atmosphere**, rivers and lakes, **soil**, glaciers and the oceans. Water is continually flowing from place to place being used and reused by plants and animals; this is represented by the **water cycle**.

The most important use of water is as a solvent. Water has polar bonds so it can dissolve ionic solids such as sodium chloride (NaCl), as well as polar solids such as **glucose** ($C_6H_{12}O_6$).

The water we drink is never pure. It always contains small amounts of gases, such as oxygen and carbon dioxide, and depending on the source of the water it may also contain dissolved solids. Some dissolved solids make the water hard. See **Hardness of water**.

Water cycle Water continually moves around the earth both in the **air** and in the oceans. Water evaporates from the oceans into the **atmosphere**. From here it falls on the earth (in several forms, such as rain, snow, etc.) and also freezes out of the air as frost and ice. This water enters the rocks and soil and will eventually find its way back into the oceans via lakes and rivers.

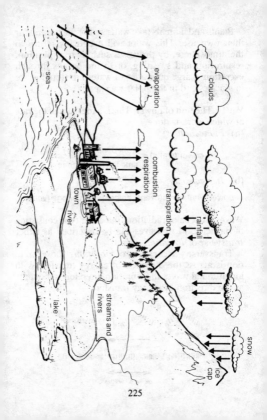

Plants and animals take water out of the ground for their own use. This water will eventually return to the atmosphere, by processes such as **transpiration**, **respiration** and sweating, or to the oceans through **excretion**. Large quantities of water are stored as ice in glaciers and in the polar ice caps.

Watt The unit of **power**. This is a measure of the rate at which work is done.

(a) *In mechanics:*

$$\text{power (watts)} = \frac{\text{work done (joules)}}{\text{time taken (seconds)}}$$

(b) *In electricity:*

$$\text{power (watts)} = \textbf{current} \text{ (amperes)} \times \textbf{voltage} \text{ (volts)}$$

Wave A type of radiation; the other type consisting of particles. Most waves can be described as either *transverse* or *longitudinal*.

Transverse waves consist of *vibrations* whose direction is across the direction of energy transfer. With longitudinal waves the vibration direction is the same as the direction of energy transfer.

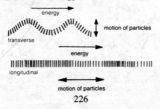

The distance taken up by a single cycle of a wave is called the *wavelength* (λ).

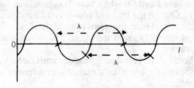

The wavelength (λ), frequency (f) and speed (c) of a wave are related in the basic wave equation;

$$\text{speed} = \text{frequency} \times \text{wavelength} \ (c = f \times \lambda)$$

All waves can show **absorption, reflection, refraction, diffraction** and interference. See **Electromagnetic waves, sound, water.**

Weight (W) This is the force which an object exerts downwards because of gravitational attraction. Like all forces, it is measured in newtons (N). The weight of an object depends on its *mass*, m, and the strength of the gravitational field, g; $W = mg$. This gives, as a unit for g, the newton per kilogram N/kg. An object's mass is thought to be constant. The value of g depends on where the object is, hence the weight of the object also depends on where it is. On the Earth ($g = 10$ m/s^2 or 10 N/kg), the weight of a 5 kg object is $5 \times 10 = 50$ N. On the Moon ($g = 1.6$ m/s^2 or 1.6 N/kg), its weight would be $5 \times 1.6 = 8$ N. In a place, such as outer space, where there is very little gravitational force (i.e. $g = 0$) acting on an object its weight will be zero. See **Gravity**.

White blood cell (or **white blood corpuscle** or **leucocyte**) One of the types of **blood cell** found in most vertebrates. Their function is to defend the body against microorganism infection. This is achieved in two ways.
(a) Some white blood cells produce **antibodies** which react with invading microorganisms and render them harmless.
(b) Some white blood cells engulf and digest invading microorganisms. This process is called *phagocytosis*. Compare **Red blood cell**.

X-rays A region of the **spectrum** of **electromagnetic waves**. The approximate **wavelength** range is 10^{-12}–10^{-10} m and the approximate **frequency** range is 10^{18}–10^{21} Hz. X-rays are produced by firing **electrons** at **metals**. It is a very penetrating **radiation** and will easily pass through flesh, but is stopped by **bone** and other dense materials. X-rays are widely used in medicine to take pictures of bones. A photographic plate is placed behind the area of the body to be investigated and the area of the body is then exposed to an X-ray source. The bones show up as light areas on the photographic plate.

Yeast A general name for a group of microorganisms which are very useful to humans. Yeasts contain **enzymes** which convert sugars into ethanol. This process is called **fermentation**.

APPENDIX A: Chemical Elements

In this Table, we give the names and symbols of the chemical elements, with their atomic numbers (Z), relative atomic masses (RAM) and melting and boiling temperatures (T_m, T_b).

Element	Z	RAM	$T_m/°C$	$T_b/°C$
actinium Ac	89	227·00	1230	3100
aluminium Al	13	26·98	660	2400
americium Am	95	243·00	1000	2600
antimony Sb	51	121·75	631	1440
argon Ar	18	39·95	−190	−186
arsenic As	33	74·92	−	610
astatine At	85	210·00	250	350
barium Ba	56	137·33	710	1600
berkelium Bk	97	247·00		
beryllium Be	4	9·01	1280	2500
bismuth Bi	83	208·98	271	1500
boron B	5	10·81	2030	3700
bromine Br	35	79·90	−7	58
cadmium Cd	48	112·41	321	767
caesium Cs	55	132·91	27	690
calcium Ca	20	40·08	850	1450
californium Cf	98	251·00		
carbon C	6	12·01	3500	3900
cerium Ce	58	140·12	804	2900
chlorine Cl	17	35·45	−101	−34
chromium Cr	24	52·00	1900	2600
cobalt Co	27	58·93	1490	2900
copper Cu	29	63·55	1080	2580

Element	Z	RAM	$T_m/°C$	$T_b/°C$
curium Cm	96	24·70	1340	
dysprosium Dy	66	162·50	1500	2300
einsteinium Es	99	254·00		
erbium Er	68	167·26	1530	2600
europium Eu	63	151·96	830	1450
fermium Fm	100	257·00		
fluorine F	9	19·00	−220	−188
francium Fr	87	223·00	30	650
gadolinium Gd	64	157·25	1320	2700
gallium Ga	31	69·74	30	2250
germanium Ge	32	72·59	960	2850
gold Au	79	196·97	1060	2660
hafnium Hf	72	178·49	2000	5300
helium He	2	4·00	−272	−269
holmium Ho	67	164·93	1500	2300
hydrogen H	1	1·01	−259	−253
indium In	49	114·82	160	2000
iodine I	53	126·90	114	183
iridium Ir	77	192·22	2440	4550
iron Fe	26	55·85	1539	2800
krypton Kr	36	83·80	−157	−153
lanthanum La	57	138·91	920	3400
lawrencium Lw	103	256·00		
lead Pb	82	207·20	327	1750
lithium Li	3	6·94	180	1330
lutetium Lu	71	174·97	1700	3300
magnesium Mg	12	24·31	650	1100
manganese Mn	25	54·94	1250	2100

Element	Z	RAM	T_m/°C	T_b/°C
mendelevium Md	101	258·00		
mercury Hg	80	200·59	−39	357
molybdenum Mo	42	95·94	2600	4600
neodymium Nd	60	144·24	1020	3100
neon Ne	10	20·18	−250	−246
neptunium Np	93	237·05	640	3900
nickel Ni	28	58·71	1450	2800
niobium Nb	41	92·91	2400	5100
nitrogen N	7	14·01	−210	−196
nobelium No	102	255·00		
osmium Os	76	190·20	3000	4600
oxygen O	8	15·99	−219	−183
palladium Pd	46	106·42	1550	3200
phosphorous P	15	30·97	44	280
platinum Pt	78	195·01	1770	3800
plutonium Pu	94	244·00	640	3500
polonium Po	84	210·00	250	960
potassium K	19	39·10	63	760
praseodymium Pr	59	140·91	930	3000
promethium Pm	61	145·00	1000	1700
protactinium Pa	91	231·04	1200	4000
radium Ra	88	226·03	700	1140
radon Rn	86	222·00	−71	−62
rhenium Re	75	186·20	3180	5600
rhodium Rh	45	102·91	1960	3700
rubidium Rb	37	85·47	39	710
ruthenium Ru	44	101·07	2300	4100
samarium Sm	62	150·40	1050	1600

Element	Z	RAM	$T_m/°C$	$T_b/°C$
scandium Sc	21	44·96	1400	2500
selenium Se	34	78·96	220	690
silicon Si	14	28·09	1410	2500
silver Ag	47	107·87	960	2200
sodium Na	11	22·99	98	880
strontium Sr	38	87·62	77	1450
sulphur S	16	32·06	119	445
tantalum Ta	73	180·95	3000	5500
technetium Tc	43	98·91	2100	4600
tellurium Te	52	127·60	450	1000
terbium Tb	65	158·93	1360	2500
thalium Tl	81	204·37	300	1460
thorium Th	90	232·04	1700	4200
thulium Tm	69	168·93	1600	2100
tin Sn	50	118·69	231	2600
titanium Ti	22	47·90	1680	3300
tungsten W	74	183·85	3380	5500
uranium U	92	238·03	1130	3800
vanadium V	23	50·94	1920	3400
xenon Xe	54	131·30	−111	−108
ytterbium Yb	70	173·04	820	1500
yttrium Y	39	88·91	1500	3000
zinc Zn	30	65·37	420	907
zirconium Zr	40	91·22	1850	4400

APPENDIX B: A list of some useful common abbreviations and symbols you may encounter in scientific literature.

A	mass number; also Ampère — unit of electric current
aq	state symbol for aqueus solution usually as (aq)
A_r	relative atomic mass
atm	atmosphere — a unit of pressure
α	alpha Greek letter
β	beta Greek letter
b.p.	boiling point
C	Celsius as in °C degree Celsius; also Coulomb — unit of electric charge (quantity of electricity)
cm^3	cubic centimetre, unit of volume
DC	direct current — the type of electricity produced from a battery
dm^3	cubic decimetre ≡ 1 litre, unit of volume
E	symbol for emf of a cell
e or e^-	electron
emf	electromotive force
g	gram — unit of mass; state symbol for gas usually as (g); also acceleration due to gravity
H	enthalpy (ΔH = enthalpy change)
I	electric current
J	Joule — unit of energy
k	prefix meaning 'one thousand times' i.e. kg = 1000 g
K	Kelvin — unit of temperature, $1K \equiv 1°C$
l	state symbol for liquid usually as (l)
m	mass; also metre — unit of length
M	molar — unit of concentration (molarity) e.g. 2M

233

m^3	cubic metre — unit of volume	
mol.	mole — unit of amount of substance	
ml	millilitre, $\frac{1}{1000}$ of 1 litre $\equiv 1$ cm^3	
m.p.	melting point	
M_r	relative molecular mass	
n	neutron	
N	Newton — unit of force	
NTP	Normal temperature and pressure	
p	proton; also pressure	
Pa	Pascal — unit of pressure	
p.d.	potential difference	
pH	relates to a scale of acidity, e.g. pH = 1 very strongly acidic	
Q	electric charge, quantity of electricity	
s	state symbol for solid usually as (s); also second — unit of time	
STP	standard temperature and pressure	
t	time	
T	temperature	
$T\frac{1}{2}$ or $t\frac{1}{2}$	half life (of radioactive isotope)	
UV	ultraviolet	
V	volume; also electrical potential difference (p.d.); also volt — unit of p.d.	
Z	atomic number	

APPENDIX C: Mathematics Symbols

Symbol	Meaning	Example
=	equals	$3 = 2 + 1$
+	add	$3 + 4 = 7$
−	subtract	$3 - 4 = -1$
×	multiply	$3 \times 4 = 12$
÷	divide	$3 \div 4 = 0.75$
$\sqrt{}$	square root	$\sqrt{9} = \pm 3$
^{n}	power	$3^2 = 9$
$\angle$	angle	$\angle$ ABC
$\triangle$	triangle	$\triangle$ ABC
<	less than	$1 < 4$
$\leqslant$	less than, or equal to	$x \leqslant y$
>	greater than	$6 > 4$
$\geqslant$	greater than, or equal to	$y \geqslant x$
$\neq$	not equal to	$4.9 \neq 49$
$\simeq$	approximately equal to	$3.9 \simeq 4$
$\propto$	proportional to	$x \propto y$
$\therefore$	therefore	

APPENDIX D: Circuit Symbols

Here we list the standard symbols and names of the main circuit elements you are likely to meet.

Conductors

	conductor
	crossing conductors
	conductor junction
	sliding contact
	simple switch
	two-way switch

Sources

—o o—	general supply to output
	cell (short arm is negative)
	battery of cells
—Ⓖ—	generator
Ψ	aerial

Passive elements

	resistor or resistive element
	fuse
	variable resistor
	capacitor
	electrolytic capacitor (black arm is cathode)
	inductor
	diode (allows current to the right)
	signal lamp
	filament lamp
	earth

Electromagnet elements

	ammeter
	voltmeter
	speaker

237

microphone

 motor

 transformer

Vacuum devices

 diode

 cathode ray tube

Semiconductor devices

 pn junction (diode)

 light-emitting diode

photo-diode

 pnp transistor

 npn transistor

 light sensitive pn diode

Sub-circuits

 gate

amplifier

239

inverter or NOT gate

AND gate

OR gate

NAND gate

NOR gate

APPENDIX E: Characteristics of Living Things

For an organism to be considered as 'living' it must demonstrate *all* of the following features:

Movement	The ability to change position either of all, or, of part of the body.
Excretion	The ability to remove from the body waste materials produced by the organism.
Respiration	The ability to release energy by the breakdown of complex chemicals.
Reproduction	The ability to produce offspring.
Irritability	The ability to sense and respond to changes in the environment.
Nutrition	The ability to take in or manufacture food that can be used when required as a source of energy or as building materials.
Growth	The ability to increase in size and complexity through the production of new cell material.

APPENDIX F: The Differences between Plants and Animals.

Plants	Animals
Cell surrounded by cellulose cell wall	No cellulose cell wall
Large vacuoles in cells filled with cell sap	Vacuoles when present only small
Large cells with definite shape	Small irregularly shaped cells
Only restricted movement possible	Free movement possible
Response to stimulus slow	Rapid response to stimulus
Cells contain chloroplasts (chlorophyll)	No chloroplasts (chlorophyll)
Photosynthesize	Must obtain food from external sources

These characteristic differences should only be regarded as guidelines. Attempts to classify certain organisms within these terms of reference will be difficult and has provided scientists with great problems, for example, Bacteria, Fungi, Viruses.

APPENDIX G: The Major Groups of Living Organisms

THE ANIMAL KINGDOM (major phyla)

(a) *INVERTEBRATES* Animals without a **vertebral column** (backbone).

Phylum Protozoa Microscopic **unicellular** animals.

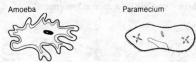

Amoeba   Paramecium

Phylum Porifera Porous animals often occurring in colonies, for example, *sponges.*

bath sponge

Phylum Coelenterata Tentacle-bearing animals with stinging cells.

Hydra   Jelly fish   Sea anenome

Phylum *Platyhelminthes* Flatworms.

Planaria

Tapeworm

Phylum *Annelida* Segmented worms.

Earthworm

Leech

Sandworm

Phylum *Mollusca* Soft-bodied animals often with shells.

Snail

Clam

Octopus

Phylum Arthropoda   Jointed limbs; exoskeleton.

Class Insecta (*louse*)   Class Crustacea *(shrimp)*

Class Arachnida (*spider*)   Class Chilopoda (*centipede*)

Phylum Echinodermata   Spiny-skinned marine animals.

Starfish   Sea urchin   Brittle star

(b) *VERTEBRATES* (Phylum *Chordata*) Animals with a vertebral column.

Class Pisces (Fish)
Fins, Scales, Aquatic.

Class Amphibia (amphibians) moist, scaleless skin, live both on land and water.

Trout

Toad

Class Reptilia (reptiles) dry scaly skin.

Class Aves (birds) feathers, constant temperature.

Turtle

Robin

Class Mammalia (mammals) Hair; constant temperature; young suckled with milk.

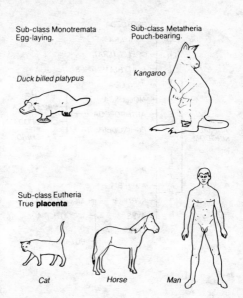

Sub-class Monotremata
Egg-laying.

Duck billed platypus

Sub-class Metatheria
Pouch-bearing.

Kangaroo

Sub-class Eutheria
True **placenta**

Cat

Horse

Man

SUMMARY

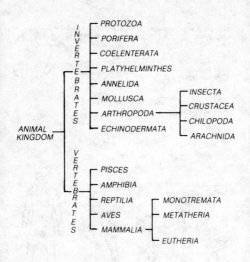

THE PLANT KINGDOM (major phyla)

Phylum Thallophyta **Unicellular** and simple multi-cellular plants.

Class Algae Photosynthetic; includes unicellular, filamentous, and multicellular types.

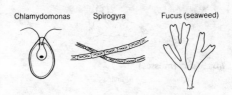

Chlamydomonas   Spirogyra   Fucus (seaweed)

Class Fungi Heterotrophic; including both parasites and saprophytes.

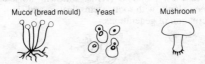

Mucor (bread mould)   Yeast   Mushroom

Phylum Bryophyta Green plants with simple leaves and showing alternation of generations; moist habitats.

Class Hepaticae (liverworts)

Pellia

Class Musci (mosses)

Funaria

Phylum Pteridophyta (ferns; bracken; horsetails) Green plants, with roots, stems, leaves, and showing alternation of generations.

Fern

Phylum Spermatophyta   Seed producing plants.

Class Gymnospermae   Seeds produced in cones.

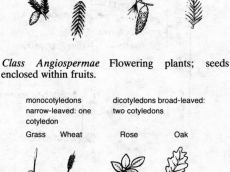

Spruce   White pine

Class Angiospermae Flowering plants; seeds enclosed within fruits.

monocotyledons
narrow-leaved: one
cotyledon

dicotyledons broad-leaved:
two cotyledons

Grass   Wheat   Rose   Oak

SUMMARY

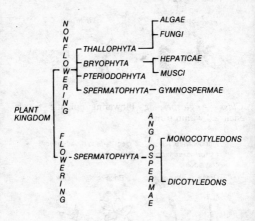

Bacteria and viruses do not meet the criteria necessary to be placed in either the animal or plant kingdoms.